"Instrumental Lives is a fabulous account of the messy lives and deaths of paper, instruments, facts and people in a physics lab in India. Sekhsaria presents an engaging story of life in a lab, and how a creative group of scientists, including himself, go about their work. But the book offers more. By working outwards from an Indian lab, the author challenges those from other parts of the world to think differently about cultures and practices of innovation."

— *Sally Wyatt,* Professor of Digital Cultures, Maastricht University

"'There can be no escaping jugaad', Pankaj Sekhsaria tells us. And in this book he beautifully traces this make-do inventiveness that marks India right into the heart of a successful nanoscience laboratory, where disaggregated parts of inkjet printers, junked computers and fridges are reassembled into a state-of-the-art scanning tunneling microscope. A compelling story of the resourceful revitalising of technologies."

— *Annemarie Mol,* University of Amsterdam

"Social studies of technology in India are few and far between. Even those few studies do not endeavour much to analyse the nuanced processes of instrument-making within a scientific organisation. The present book makes a noteworthy contribution in this regard. It showcases the relevance of 'jugaad' in the mainstream scientific research, thereby, making a welcome departure from many other studies in the contemporary discourse on technology and innovation."

— *Saradindu Bhaduri,* Centre for Studies in Science Policy, Jawaharlal Nehru University, New Delhi

"This fascinating ethnographic work explains the various meanings of innovation(s) and the significance of taking geographies of innovation seriously to understand how a technology/instrument is used and re-used. By looking at *Jugaad*, Pankaj Sekhsaria raises important questions about the ways in which what he calls as *Technological Jugaad* is created, welcomed, and rejected in a laboratory. This well-written ethnography of a laboratory will be an important contribution in Science and Technology Studies and History of Science."

— *Renny Thomas,* Assistant Professor of Sociology, Jesus and Mary College, University of Delhi

Instrumental Lives

Instrumental Lives is an account of instrument making at the cutting edge of contemporary science and technology in a modern Indian scientific laboratory. For a period of roughly two-and-half decades, starting in the late 1980s, a research group headed by CV Dharmadhikari in the physics department at the Savitribai Phule University, Pune, fabricated a range of scanning tunnelling and scanning force microscopes including the earliest such microscopes made in the country. Not only were these instruments made entirely in-house, research done using them was published in the world's leading peer reviewed journals, and students who made and trained on them went on to become top class scientists in premier institutions.

The book uses qualitative research methods such as open-ended interviews, historical analysis and laboratory ethnography that are standard in Science and Technology Studies (STS), to present the micro-details of this instrument making enterprise, the counter-intuitive methods employed, and the unexpected material, human and intellectual resources that were mobilised in the process. It locates scientific research and innovation within the social, political and cultural context of a laboratory's physical location and asks important questions of the dominant narratives of innovation that remain fixated on quantitative metrics of publishing, patenting and generating commerce.

The book is a story as much of the lives of instruments and their deaths as it is of the instrumentalities that make those lives possible and allow them to live on, even if with a rather precarious existence.

Pankaj Sekhsaria was until recently Senior Project Scientist at the DST-Centre for Policy Research, Department of Humanities and Social Science, IIT-Delhi. His research interests lie at the intersection of science, environment, technology and society. He has a PhD in Science and Technology Studies (STS) from the Maastricht University in the Netherlands and has written extensively on issues of environment, development and wildlife conservation in India with a special focus on the Andaman and Nicobar Islands.

Routledge Focus on Modern Subjects
Series Editor: Saurabh Dube
Research Professor of History, El Colegio de México, Mexico City

Routledge Focus on Modern Subjects has a broad yet particular purpose. It explores quotidian claims made on the *modern* – understood as idea and image, practice and procedure – as part of everyday articulations of modernity on the Indian sub-continent. Here, the category-entity of the *subject* has wide purchase. It refers not only to social actors who have been active participants in historical processes of modernity, but equally implies branch of learning and area of study, topic and theme, question and matter, and issue and business.

The series addresses such modern subjects in a range of distinct yet overlaying ways, focusing on capital and class, culture and power, gender and identity, politics and privilege, nation and narration, design and dominance, aesthetics and authority, and science and subjectivities – in everyday and institutional arenas.

Instrumental Lives
An Intimate Biography of an Indian Laboratory
Pankaj Sekhsaria

For more information about this series, please visit: www.routledge.com/Routledge-Focus-on-Modern-Subjects/book-series/RFOMS

Instrumental Lives

An Intimate Biography of an Indian Laboratory

Pankaj Sekhsaria

Routledge
Taylor & Francis Group

LONDON AND NEW YORK

First published in paperback 2024

First published 2019
by Routledge
4 Park Square, Milton Park, Abingdon, Oxon OX14 4RN

and by Routledge
605 Third Avenue, New York, NY 10158

Routledge is an imprint of the Taylor & Francis Group, an informa business

© 2019, 2024 Pankaj Sekhsaria

Publisher's Note
The publisher has gone to great lengths to ensure the quality of this reprint
but points out that some imperfections in the original copies may be
apparent.

British Library Cataloguing-in-Publication Data
A catalogue record for this book is available from the British Library

Library of Congress Cataloging-in-Publication Data
A catalogue record for this book has been requested

ISBN: 978-1-138-58767-0 (hbk)
ISBN: 978-1-03-293073-2 (pbk)
ISBN: 978-0-429-44965-9 (ebk)

DOI: 10.4324/9780429449659

Typeset in Times New Roman
by Apex CoVantage, LLC

This one is for my little Kabir, eight years old now, and a bundle of joy and energy.

Contents

Illustrations

Series editor's foreword

Saurabh Dube

It is a pleasure to write this foreword to the first title in the *Routledge Focus on Modern Subjects* series. In what follows, I shall first introduce the series and then turn to the book.

The series

Routledge Focus on Modern Subjects has a broad yet particular purpose. It seeks to explore quotidian claims made on the *modern* – understood as idea and image, practice and procedure – as part of everyday articulations of modernity on the Indian sub-continent. Here, the category-entity of the *subject* also has wide purchase. It refers not only to social actors who have been active participants in historical processes of modernity, but equally implies a branch of learning and area of study, topic and theme, question and matter, and issue and business. The series attempts to address such modern subjects in a range of distinct yet overlaying ways.

Questions of modernity have always been bound to issues of being/becoming modern. These themes have been discussed in various ways for a long time now.[1] For convenience, we might distinguish between two broad, opposed tendencies. On the one hand, over the past few centuries, it is the West/Europe that has been seen as the locus and the habitus of the modern and modernity. Such a West is imaginary yet tangible, principally envisioned in the image of the North Atlantic world. And it is from these arenas that modernity and the modern appear as spreading outwards to transform other, distant and marginal, peoples in the mold and the wake of the West. On the other hand, such propositions have been contested by rival claims, including especially from within Romanticist and anti-modernist dispositions. Here, if the modern and modernity have been often understood as intimating the fundamental fall of humanity, everywhere, so too have the aggrandisements of an analytical reason been countered through procedures of a hermeneutic provenance.

Needless to say, these contending tendencies have for long found imaginative articulations, and I provide indicative examples from our own times. The work of philosophers such as Jürgen Habermas and Charles Taylor and historians such as Reinhart Koselleck and Hans Ulrich Gumbrecht have opened up the exact terms, textures and transformations of modernity and the modern. At the same time, they have arguably located the constitutive conditions of these phenomenon in Western Europe and Euro-America. In contrast, anti-modernist sensibilities have found innovative elaborations in, say, the "critical traditionalism" of Ashis Nandy in South Asia; and the querying of Eurocentric thought has been intriguingly expressed by the scholars of the "coloniality of knowledge" and "decoloniality of power" in Latin America. These powerful positions variously rest on assumptions of innocence before and outside Europe and the West, modernity and the modern.

Engaging with yet going beyond such prior emphases, recent work on modernity has charted new directions, departures that have served to foreground questions of modernity in academic agendas and on intellectual horizons, more broadly. I indicate four critical trends. First and foremost, there have been works focusing on different expressions of the modern and distinct articulations of modernity as historically grounded and/or culturally expressed articulations that query *a priori* projections and sociological formalisms underpinning the category-entity. Second, there are the studies that have diversely explored issues of "early" and "colonial" and "multiple" and "alternative" modernity/modernities. Third, we find imaginative ethnographic, historical and theoretical explorations of modernity's conceptual cognates such as globalisation, capitalism and cosmopolitanism as well as of attendant issues of state, nation and democracy. Fourth and finally, there have been varied explorations of the enchantments of modernity and of the magic of the modern, understood not as analytical errors but as formative of social worlds. These studies have ranged from the elaborations of the fetish of the state, the sacred character of modern sovereignty, the uncanny of capitalism, and the routine enticements of modernity through to the secular magic of representational practices such as entertainment shows, cinema and advertising.

Routledge Focus on Modern Subjects engages and exceeds, takes forward and departs from such concerns in its own manner. To start off, its titles address the queries and concepts entailed in earlier explorations of the modern and recent reconsiderations of modernity by focusing on a clutch of common and critical questions. These issues turn on the everyday elaborations of the modern, the quotidian configurations of modernity, on the Indian sub-continent. Next, rather than simply asserting the empirical plurality of modernity and the modern, the cluster approaches the routine, even banal, expressions of the modern as registering contingency, contradiction and contention as lying at the core of modernity. Further, it only follows that our bid is not to indolently exorcise aggrandising representations of modernity *as* the West, but to prudently track instead the play of such projections in the commonplace unravelling of the modern in India today. Finally, such procedures not only recast broad questions – for instance of cosmopolitanism and globalisation, state and citizenship, Eurocentrism and Nativism, aesthetics and authority – by approaching them through routine renderings of the modern in contemporary South Asia. They also stay with the dense, exact expressions of modernity yet all the while attending to their larger, critical implications, prudently thinking *both* down to the ground.

In keeping with the spirit of the series, all its titles stand informed by specific renderings – as well as focused rethinking – of key categories and processes. Let me provide two exact instances. In different ways, concepts and processes of power and politics alongside those of community and identity variously run through the *Routledge Focus on Modern Subjects* series. Here, neither power nor politics are rendered as signifying solely institutional relations of authority centering on the state and its subjects. Rather, the bid is to articulate these as equally embodying diffuse domains and intimate arrangements of authority and desire, including their seductions and subversions. Actually, as parts of such force-fields, state and government, their policy and program might now assume twinned dimensions in understandings of modern subjects. Here can be found densely embodied disciplinary techniques towards forming and transforming subjects-citizens, where such protocols and their reworking by citizens-subjects no less register the shaping of authority by anxiety, uncertainty and alterity, and/ or of the structuring of command by deferral, difference and displacement.

At the same time, the series approaches community and identity as modern processes of meaning and authority, located at core of nation and globalisation. This is to say that instead of approaching identity and community as already given entities that are principally antithetical to modernity, this cluster explores communities and identities as wide-ranging processes of formations of subjects, expressing collective groupings and particular personhoods. Defined within social relationships of production and reproduction, appropriation and approbation, and power and difference, emergent identities, cultural communities, and their mutations appear now as essential elements in the quotidian constitution, expressions and transformations of modern subjects.

The book

Instrumental Lives articulates, indeed embodies, the concerns of this series in many, apposite ways. Located in the field of Science and Technology

Studies (STS), turning on methods of "lab ethnography" and the theory of "social construction of technology," the book construes an "intimate biography" of a laboratory in the physics department of the Savitribai Phule University, Pune, in western India. Pankaj Sekhsaria focuses on the making of scanning tunneling microscopes (STMs), central to nanotechnology research, by Professor CV Dharmadhikari and his students, in order to track the lives of these instruments – and the space(s)-time(s) they have inhabited – from their fledgling beginnings through to their touching ends. At stake are inventions and innovations – as well as disappointments and dead-ends – that bring into *play* everyday objects, vocational sciences and technological creations.

All of this allows the author to layer and un-layer the terms and horizons of *jugaad*, further underscoring the salience of technological jugaad. Rescued from the condescension of reductive, adjudicatory apprehensions of the phenomena, in Sekhsaria's hands, jugaad intimates a method and manner that is grounded in Indian imaginaries yet one that bears wide associations, critical implications. Indeed, as articulated in the work, technological jugaad militates against instrumentalist understandings of science and technology, development and progress, often intimately associated with imperatives of state and interests of capital. In their stead, Sekhsaria brings into view inherently democratic, avowedly anti-instrumentalist, imaginatively inclusive orientations to citizenry and skill, waste and technology, everyday knowledges and vocational sciences.[2]

As part of this series, there are distinct registers of *modern subjects* that *Instrumental Lives* at once operates upon and innovatively animates. To begin with, there are the scientists themselves, not only Dharmadhikari, his students and their ilk but other practitioners too, these varied sets of subjects bearing distinct, even opposed, orientations to science and technology, innovation and knowledge-making. Here are to be found modern subjects whose meaningful practices in the lab and discursive dispositions at large reveal that there are different ways of being modern. Alongside, ahead of us are also social subjects whose exact divisions yet uncertain overlaps no less register the fault-lines and ambivalences at the core of modern regimes of disciplinary knowledges, which are part of the contentions and contradictions of modernity.

It only follows that the instruments and the lab explored by the book equally announce modern subjects, now in the sense of a branch of learning and area of study, topic and theme, question and matter, and issue and business. Yet, it warrants emphasis that the instruments and the laboratory are far from being simply inert objects of a disinterested science. Instead, they acutely insinuate discrete subjects that acquire lives and meanings in the social labour, the relational work, of knowledge making. In as much as there are distinct dispositions in the fields of science and technology to instruments and labs, these subjects not only find alternative articulations in scientific research as vocational practice, they also appear as variously challenging yet unsurprisingly entangled with the legislative propensities of modern knowledge, and its adjudicatory reason. Together, on offer once more are the contingencies and conflicts of modernity.

There are other reasons, too, for the close connections between the crucial attributes of this title and the primary purposes of the *Routledge Focus on Modern Subjects* series. On the one hand, the series is premised upon the importance of attending to – what Michel de Certeau has called – the *details*, in this case concerning the unravelling of the modern and modernity on the Indian subcontinent. In contrast to immaculate imaginings of the hyper-organised site of the modern laboratory, Sekhsaria brings to life the constitutive clutter that contains the effective order and logic to be found in the apparent disorder and disarray in the spaces it studies. This is in keeping with the careful attention paid in *Instrumental Lives* to the making of STMs, fabricated out of bits and pieces, from an abandoned refrigerator shell through to other routinely everyday objects. Although such a seemingly haphazard practice of science appears unconnected to the overwrought machines of scientistic schemas, modernist reifications and technocratic desires, which can all poignantly characterise also quotidian conceptions, it actually stands at the core of labs and their logics as the field of STS – its attendant analyses, emphases and ethnographies – has been at pains to tell us.

On the other hand, in Sekhsaria's hands critical understanding is accompanied by careful considerations, such that analytical critique, ethical concern, and epistemological caution go hand in hand. This is in keeping with the spirit of the series, which eschews the turning of power – for instance, of state and nation, modernity and capital, science and technology – into a readily distant enemy, an endlessly dystopic totality, while celebrating difference – for example, of the subaltern, the everyday, or in indeed "alternative science" – as an easy antidote to power, a principally "un-recuperated particular." Thus, the work modestly uncovers its own conditions of possibility as inherent in studies of science that have queried the privileging of "expert knowledge" and avowed instead the place of wider "social movements" in this scenario. Yet, Sekhsaria also distances his method and intent from such studies, prudently pointing to how ethnographic procedures – rather than textualist and archival readings alone – better unravel the detailed textures of academic arenas of technology and science. Similarly, rather than readily railing against an exclusive enemy, out-there, somewhere, *Instrumental Lives* attends to the ambiguities and ambivalences that characterise the problems and possibilities of science and technology in the present. At stake are acute mix-ups marked by distinct modalities of knowledge-making and claims of capital, technocratic blueprints and statist desires, instrumentalist knowledges and vocational sciences.

Taken together, these twin attributes of the study – a field-based hermeneutics of the terms and textures, resonances and dissonances, details and designs of labour in/of a laboratory *alongside* the task of thinking through and staying with the opacities and opportunities of/in the present concerning S&T – allow Sekhsaria to track the making and unmaking of instruments and sciences within the force-fields of disciplinary knowledge and political economy, subject and state, which come together yet also fall apart. Such measures allow *Instrumental Lives* to further point towards wider sets of questions and concerns. It is to four such issues that I now briefly turn.

As part of the broader focus and purpose of STS, *Instrumental Lives* pays close attention to instruments and their innovation, labs and their life as located in particular places, specific spaces. This is to say that rather than consider science and technology as announcing the normatively neutral view from nowhere that turns into the inherently transcendental vista for everywhere, unconstrained by time and space, Sekhsaria articulates how Dharmadhikari became part of an international "instrumental community" of probe microscopists, a community that was co-created over time and where the scientists were at once "technologists and community builders." Similarly, discussing the making of STMs in Dharmadhikari's laboratory, the work imaginatively elaborates the "local geographies" of Pune – entailing, for instance, intimate knowledge of markets, materials and (skilled) machinists, as well as the presence of experts, engineers and (educational) institutions in the city – such that quotidian cartographies emerge as crucial to the production of scientific knowledge and technological innovation.

Now, such emphases on specific space(s) "as constitutive of systems of human interaction" and on the salience of situated sites in the fabrication of critical instruments is valuable precisely because they do not approach space as a passive context, abstracting practices from the worlds in which they are embedded. At the same time, it is worth enquiring if, beyond the impact of space in shaping human action, at stake in these instances is not the *actual production of space and time* through the practices and meanings of social subjects? That is, I am asking if the creation of the wider "instrumental community" of probe microscopists and of the local fashioning of STMs did not itself involve the varied construal and co-constitution of space and time of the scientific community, technological practice, laboratory activity and a particular Pune within and through the imaginaries and activities of social subjects – the technologists and scientists as well as their accomplices and co-conspirators?[3]

To raise such questions is to widen the range of address of *Instrumental Lives*. It is to register that STS importantly points towards a refusal to reify labs and their instruments, science and technology as bracketed from social arenas. Yet, it is equally to enquire: How are we to further situate these expert practices as part of different domains of routine life, as such? This is to ask if it is perhaps the case that STS remains somewhat wedded to its object of critique and subject of desire, rather than displace and disinter – while exploring and interring – these within the larger force-fields of every-day and institutional practices, human and non-human arenas?

It follows that ahead of Sekhsaria's able unravelling, across the book, of dominant projections of S&T as well as of cultures of innovation – the former especially enmeshed with developmental regimes of state, nation, and capital; the latter formidably bearing a Schumpeterian provenance – I was left uncertain by a leap of faith on his part: "Alternative narratives are possible – they already exist, in fact. All that is needed is for us to be sensitive to their possibilities and to be open to what they might have to offer." On the one hand, in our murky worlds, can these alternatives really embody such innocence? Is sensitivity and openness to their possibilities actually "all that is needed"? Have we not returned, now from the perspective of margins, to the guarantee of progress that heroically upholds the warranty of history under regimes of modernity? Concerning the discussion in the book, were the structures that gave birth to "encultured modalities of innovation" – and sustained their subsequent lives – not intimately bound also to pro-cesses that begat the demise of innovative instruments as well as the death of the lab that produced them? On the other hand, do not such assertions overlook the immense burden of an instrumentalist, meaning-legislative, totalising reason in authoritative schemas of power and knowledge, which variously (over)rule the world? What is at stake in eschewing easy alter-natives in order to carefully construe instead hermeneutic understandings, ethical explorations and melancholy realisms, which are open to revising their deeply held constitutive presuppositions while being ever suspicious of adjudicatory reason?[4]

At the end, as we ponder the superannuation of a scientist (Dharmadhi-kari), the closure of his lab and the abandoning of instruments fashioned there, the lumps in our throats remind us that the purpose of knowledge is better served by dispositions towards caring to know, rather than aggrandis-ing, muscular analytics claiming secular transcendence. Here, the instru-ments that inhabit this book far from being mere objects turn into living forms, bearing "value properties" that make calls to action and apprehen-sion in human worlds, in relation to the scientists who created them – the instruments and their makers both intimating modern subjects, affective subjects of a worldly immanence.

Notes

1 The discussion in this Foreword of different understandings of modernity (and the modern) draws upon a wide range of scholarship. Instead of cluttering the short piece with numerous references, let me only indicate a few of my works that have addressed these themes – in dialogue with relevant literatures – and that back my claims here. Needless to say, I am cryptically condensing and radically rearranging my prior arguments and emphases for the present purposes. Saurabh Dube, *Subjects of modernity: Time/space, disciplines, margins* (Manchester: Manchester University Press, 2017); Saurabh Dube, *Stitches on time: Colonial textures and postcolonial tangles* (Durham and London: Duke University Press, 2004); and Saurabh Dube, *After conversion: Cultural histories of modern India* (New Delhi: Yoda Press, 2010). Consider also, Saurabh Dube (Ed.), *Enchantments of modernity: Empire, nation, globalization* (New Delhi: Routledge, 2009); and Saurabh Dube (Ed.), *Handbook of modernity in South Asia: Modern makeovers* (New Delhi: Oxford University Press, 2011).

2 This is particularly the case with the compelling critique offered in the book of Science and Technology (S&T) as destiny and horizon that is tied to regimes of capital and state (fueled into the "enterprising lab") today, including the haunting of vocational sciences by such schemas. It should also be clear that my discussion of the querying by Sekhsaria of instrumental reason avows yet works upon distinct registers of *instrumentality* and *instrumentalization*. It bears emphasis that, having traced the overlaps and distinctions of jugaad with cognate conceptions such as those of bricolage, here is how Sekhsaria sums up his discussion of the subject, finding "technological jugaad as something explicitly Indian because of its location in a culture of languages across northern India and the broad sweep of its canvas from rural agricultural and poor urban India on the one hand to the modern scientific laboratory on the other, which considers the context of economic and resource constraints that circumscribe life in large parts of the country, including in the laboratory – from a particular understanding of waste that allows for the reconfiguring of materiality to a contingent culture of recycling that feeds and is fed by the imperatives of survival and also to the existence of alternative, often traditional, systems of knowledge."

3 See especially, Dube, *Subjects of modernity*.

4 Taken together, in several of the writings cited previously, I elaborate such procedures of careful probing and critical affirmation of social worlds and their knowledges, grouping the protocols as a "history without warranty."

Acknowledgements

Putting together this book would not have been possible without the help, support and good wishes of a number of people and I am extremely grateful to all of them. I would like to thank my doctoral supervision team at Maastricht University – first and foremost, my supervisor Wiebe Bijker – a better friend, philosopher and guide I could not have asked for; Aalok Khandekar for his incisive questions, sharp conceptual clarity and deep engagement; Ragna Zeiss for her constant interest and support; and also to Koen Beumer, Trust Saidi and Charity Urama, the other doctoral candidates in the project.

Others at the Faculty of Arts and Social Sciences (FASoS) at Maastricht University whom I had the opportunity to interact with and learn from included Karin Bijsterveld, Geert Somsen, Jo Wachelder and Esha Shah. I would also like to thank the entire class of the CAST Masters of 2011 – Ranjit Singh, Bart Zwegers, Ana Teresa Pires, Aline Reichow, Rick Holsgens, Trust Saidi, Annapurna Mamidipudi, Jeremias Herberg, Charity Urama, Tuur Ghys, Megan Curtis and Brian Keller. Special thanks are due to Ranjit Singh and to Samir Passi for their constant and unconditional company and for the endless conversations I've had with them.

Many thanks also to Willem Halfman and Teun Zuiderent-Jerak, the coordinators of the WTMC workshops and summer schools and the many co-doctoral candidates I had the opportunity of meeting, talking to and learning from during this period.

In India, of course, the research and the book would not have been possible without the openness and interest showed by CV Dharmadhikari and the other researchers who made up this research group – Subhendu Dey, SB Iyyer, Sadhu Kolekar, Rajendra Kshirsagar, Shivaprasad Patil and Sumati Patil. In allowing me into their labs, their research and their lives, they showed huge trust and faith and I am deeply indebted to them for this. I would also like to thank Arun Nigvekar, Vijaymohanan Pillai, Arup Raychaudhuri and Murali Sastry for the time and the interviews they gave me and the many insights they offered.

xxii *Acknowledgements*

My deep gratitude also goes to my parents and brother Peeyush. And to
Latha, without whose willingness, companionship, patience and support,
my doctoral research and this book would never have come into being.

For the last two years I've been at the DST Centre for Policy Research
in the Department of Humanities and Social Science, Indian Institute of
Technology Delhi, New Delhi. I am grateful for the opportunity the centre
has offered me and to my colleagues there, Naveen Thayyil in particular,
for the constant support. Many thanks also to the anonymous reviewer for
their very encouraging words and detailed comments that helped hugely in
making this a much better manuscript. Thanks also to Aafreen Ayub, my
editor at Taylor & Francis, for being at the helm right through and taking us
all smoothly through the many steps a book needs to go through before it
reaches the reader's hands.

Last but not the least to Saurabh Dube, whom I met completely by chance
on the sidelines of the Goa Arts and Literature Festival in December 2017.
This book is not an outcome of that meeting . . . that meeting was the purpose of this book. And there's only one way to describe it all – serendipity!

1 Introduction

Entering the lab/setting the stage

It was in a moment of desperation in early December 2010 that I had called up the telephone number of the physics department at the Savitribai Phule University in the city of Pune. I was trying to get in touch with Prof CV Dharmadhikari, Head of the Center for Advanced studies in Materials Science and Solid State Physics at the department here. It was nearly a year since I had begun my PhD work on "The Cultures of Innovation in nanotechnology research for development in India," and I had made little progress in identifying laboratories that would agree to be the subjects of my study. With one partial exception, I had no case-study or even a potential case-study to report on and this was a situation that needed urgent correction.

Surfing the internet a few days earlier in an attempt to find who was researching the world of nano in the city of Pune, my attention had been caught by the sparse web page of Dharmadhikari's research group on the website of this physics department. The research interests of the group included "Development and application of the Scanning Tunneling Microscope and Atomic Force Microscope" and "Development of special techniques for surface diffusion measurements using tunneling and field emission current fluctuations."[1] At first glance this seemed extremely interesting and potentially a very good case if things would work out. I immediately emailed Dharmadhikari, briefly outlining my research project and requesting for an opportunity to meet him. I received no response at all.

1.1 Meeting the scientist

In normal course I would have sent another email reminder and waited a while more to hear from the professor; I have increasingly begun to prefer the lack of immediacy offered by email communication as against the immediate call to action that a phone call often elicits or entails. But there are exceptions and this was one of them.

The operator put me through to the relevant extension, which was answered by a female voice I reflexively referred to as 'ma'am.' I introduced myself, explained the context of my call and asked if I could speak to Prof Dharmadhikari. I was told he was not in the lab and I asked tentatively if I could get the number of his mobile phone; tentatively, because I did not expect to easily get the number of a senior scientist who must surely be an extremely busy man. Quite to my surprise then, she gave me the number without any hesitation and also suggested that I could call him straight away if I wanted to.

This I did, though I was not very sure of what to expect from the other end of the line. I introduced myself and to my immense relief and pleasant surprise, was immediately recognised by the professor. Yes, his friendly voice said, he'd seen my email and he would be happy to meet me. "Can I come over, right away?" I pushed my luck. An affirmative answer again – yes, I could. It seemed like a good day and I hopped immediately into an auto rickshaw and headed to the university. I asked directions to his room, only to find it bolted. I buzzed his mobile again, this time with a lot less hesitation. The tone of his voice indicated he was surprised that I had already arrived, but he was still very friendly. He was having tea at the department canteen and invited me to join him there.

I headed out to the back of the building and walked passed the badminton court to the open-air canteen where two senior men sat on plastic chairs at a large wooden table. One of them, I presumed rightly, was the person I had come to meet. I asked for Prof Dharmadhikari, introduced myself and handed my university business card to both the gentlemen. Dharmadhikari quickly took charge of the situation. He asked me to sit down, offered me a cup of tea and introduced me to his colleague as a social scientist who had come to study some of the STM related work that he had been doing. As I sat and sipped my tea in silence (my silence) I caught some snatches of the conservation the two continued to have – mainly gossip I would say – about what is happening in the department, about the work of some other scientists and random bits of this and that. I waited, hoping for their conversation to wind-up soon so that my work could begin. Just at it seemed to be ending, however, the canteen space was caught up in an unexpected flurry of activity that had them occupied all over again. A very senior (and also senior-looking) professor walked in followed by a number of busy and officious looking persons including other senior professors and members of staff.

This, I found out subsequently, was Prof Ram Takawale who was not only a former head of this physics department, but had also been the vice chancellor of the university. The attention, presumably, was well deserved, the flurry of activity and attention, well justified. He and his retinue sat down at the adjoining table and the two professors I was with also joined the ensuing conversation that meandered from present developments in the university, to the situation in the department to a discussion of a particular professor in a particular department to, as it often happens in discussions such as these, the state of Indian politics. There was good humoured banter and a round of tea was ordered with one steaming cup coming my apprehensive way as well. I hope this gets over soon, I kept telling myself.

Eventually, and actually not much later, it did indeed finish. Takawale and his retinue departed and Dharmadhikari suggested we go his room. I wished his companion professor good-bye and followed Dharmadhikari past the badminton court again, up a small flight of steps and into a dimly lit corridor with a row of doors on the right that had wooden name boards hung on top of the wooden frames.

Dharmadhikari's was a small room, much smaller than I had expected. There was a table with a chair on the other side that Dharmadhikari occupied. On this side, the side towards the door, were two other chairs. As I sat down in one of them, I looked around. This was a cubicle about 10x10 feet in dimension, maybe a little larger, and the overarching memory I carry of this space is of paper. Along the walls on my right and my left were waist height cabinets placed on the floor with wall cabinets above. The top of these cabinets and the shelves inside were full of paper – loose sheets, stapled sets, files, books – some piled one on top of the other, some in stacks that stood neatly and others that had scattered and slipped like happens when one set is pulled out in a hurry from the middle of a pile. Even the table that we were sitting at had only a small patch where the top was visible. The rest was all covered by paper. The only other thing I remember from that room was a computer and a printer that sat by the side.

As we settled down, Dharmadhikari looked at my business card again and asked me about my research. For the next few minutes he listened intently as I tried my hand at explaining what I was trying to do – the field of Science and Technology Studies (STS) that was the framework for my research project, innovation in nanotechnology research in India, lab ethnography, the theory of the Social Construction of Technology – SCOT, my supervisor's work on the bicycle (Bijker, 1995b; Pinch & Bijker, 1987),[2] the value of such studies of science and technology (S&T), issues around the democratisation of S&T, the history of the scanning tunneling microscope (STM) and a bit more that I had picked up in the preceding few months of studying STS and exploring the field of nanotechnology itself.

He seemed generally agreeable and I thought it was a good chance to ask formally for an interview that I could record. I asked tentatively and his first reaction was of hesitation, if not outright denial. I pushed a bit, assuring him that I wouldn't use anything without his knowledge and there would be nothing for him to worry about. He gave his nod eventually but it was clear it had been given with reluctance. He attributed his resistance to a very unpleasant experience many years ago, in 1998, when the news daily, *The Indian Express*, had carried a report on the development of his in-house STM:

> Can you imagine the [University's] Jayakar Library's entire collection of over 4 lakh [four hundred thousand] books on a floppy or for that matter an elephant the size of an ant? Apparently not. But the development of the country's first ever indigenous state-of-the-art scanning tunneling microscope (STM) at the Department of Physics, University of Pune, may well make the fairy tale come true, as not only does it take a fine probe of how a diamond grows – layer upon layer – but can also manoeuvre genes at an atomic level!
>
> (Shah, 1998)

The elephant to ant illustration, Dharmadhikari said, he had used only as a "light hearted" analogy in his conversation with the journalist. The way it was reported, however, and the added allusion to "mavoeuver[ing] genes at an atomic level" had conveyed a completely different picture. All kinds of queries came his way and many explanations were demanded, including by some "human rights fellows," as Dharmadhikari referred to them. I didn't figure out how the issue was resolved, but the episode had certainly made Dharmadhikari wary of people walking around with recording devices and asking for interviews.

1.2 Entering the labs

Later that afternoon Dharmadhikari took me briefly to his two (physically) small labs – one the scanning tunneling microscopy (STM) lab, the other his scanning force microscopy (SFM) lab (Image 1.1). I met two of his latest doctoral students there, was shown a power point of some of the work that had been done in the labs over the years and also the images they had made with their instruments. The most striking of these for me was the nano-sized $(200 \text{ Å} \times 200 \text{ Å})^3$ image of the syllable ॐ etched out in the Devnagiri script on a surface of gold. I found this image striking for two reasons – first, because it came across a bit like repeating the iconic 1990 image of IBM that had been spelt out with 35 individual xenon atoms (Baird, 2004a; Eigler & Schweizer, 1990)[4] and second (retrospectively), because of the significance that has as a religious symbol and in the stories of origin in Hindu religion and mythology.

I experienced the labs themselves as hugely scattered clutter – small spaces that were crammed with chairs, computers, books shelves, cupboards and tables on which one saw a number of things: tools such as pliers, screw

Image 1.1 The scanning force microscopy laboratory
Source: Author

drivers, nuts, bolts, small boxes of plastic and aluminium, double sided tape, glue sticks, scraps of paper, sheets of paper, files, books, pens, pencils, circuit boards, streams of wires running from here to there and, of course, a series of big and small instruments (Images 1.2 and 1.3; also see Annexure 2). In the corner of one lab was a refrigerator shell (Image 5.1) and a plywood box (Image 5.5), each housing a STM that had been fabricated in this lab. In the other lab I saw a large upturned aluminium vessel of the kind we use to boil water on the gas stove in the kitchen at home. Underneath was a contraption that lay on the inflated tube of an automobile tyre – another STM (Image 5.3), which too had been made in-house.

At first sight this set-up did not conform at all to my (and presumably also to many non-scientist's) general idea of labs as clean, organised and disciplined spaces (Traweek, 1988, pp. 56–57). It was, in fact, the archetype and the material embodiment of disorder from which the scientific enterprise produces order and meaning (Latour & Woolgar, 1986; Traweek, 1988).

I can say this with confidence now only on account of much retrospective insight, but, even then, a lot of it seemed counter-intuitive. It was in many ways the first indication that there was something here that went

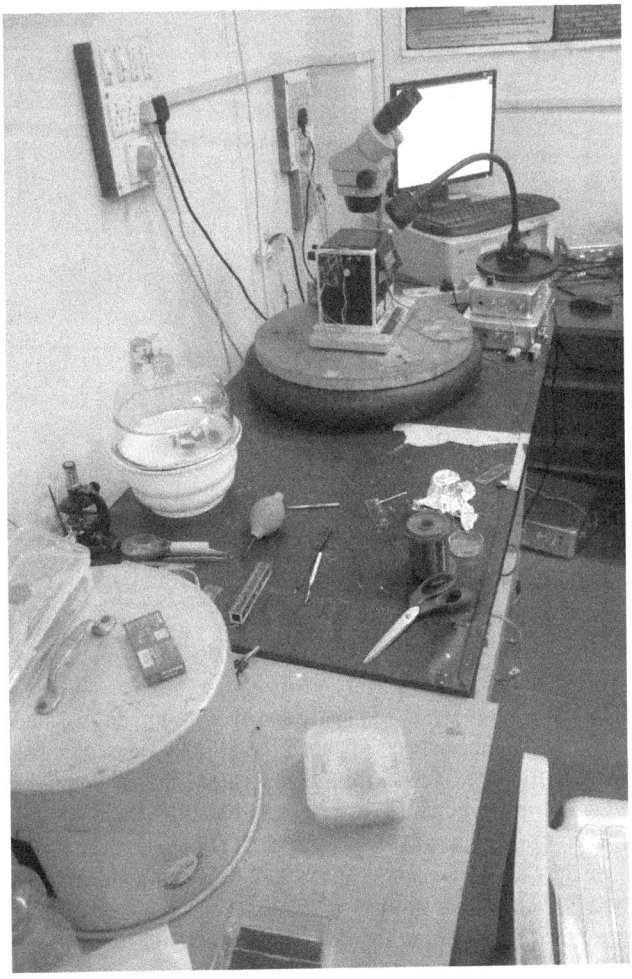

Image 1.2 A workbench in the lab

Source: Author

much deeper than was visible on the surface (see Annexure 1 for my diary notes from this first meeting with Dharmadhikari). There was a sense of excitement and anticipation of what my research would reveal and what the eventual story would be. But more than anything else, it was a sense of immediate relief I felt at that moment – I had managed to get access to the labs and my 'work' could finally begin!

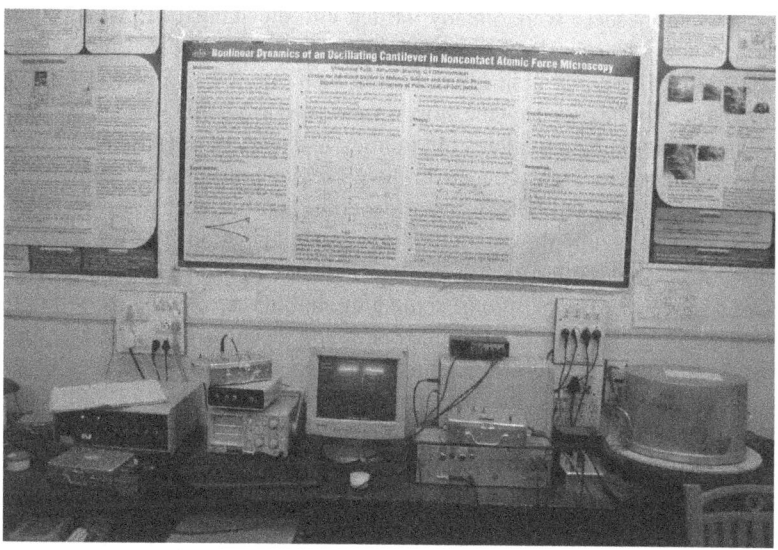

Image 1.3 A workbench in the lab
Source: Author

1.3 The first set of insights

The short, 30-minute interview that I conducted with Dharmadhikari and the brief tour of the labs that first day laid out, very broadly and in an unstructured manner, the history, trajectory and chronology of Dharmadhikari's scientific journey that included the indigenous and in-house creation of among the earliest, if not *the* first STMs and atomic force microscopes (AFMs) in India. This was a journey that started somewhere in the mid-1980s (though one can pull it back further into time) and to which my doctoral research became an unexpected and unplanned add-on rather quickly. It helped that Dharmadhikari was himself intrigued about what I was doing – just the fact that someone was interested in the history of his scientific endeavours and also that something like STS existed at all.

Somewhere towards the end of that first day Dharmadhikari told me that under normal circumstances he would not have given the time he had given me. He had decided to make an exception following our phone conversation, he explained, because he had sensed excitement and was interested in the subject himself. Looking back retrospectively, the decision he made that day to give me access and also his time over the next many months had a much larger significance than I had initially assumed. It is conjecture on my part today, but these were already written into the subtext of that specific context and what unfolded because Dharmadhikari had retired just a few days before I first met him. I had thought then that it was a great time for me to get into the picture – the story I would tell would be like the perfect closing of the loop – an interesting journey in a modern physics laboratory now coming to its logical end with the professor retiring after many decades of work. As I was to realise in the months that followed, however, this perfect loop was only the creation of my own imagination. There was/is no perfect loop, and there surely is no perfect closure. The point I entered may have seemed like the closure of something, but it was also the entry point for something else – something like one funnel opening into another – the first one moving towards closure but actually opening out into another huge world that could funnel into unknown directions.

My research was not structured in any particular manner. It moved unexpectedly like any research does – sometimes smooth and fast, sometimes stagnant, sometimes turbulent eddies; not threatening in any way, but eddies nevertheless. There were a number of things I did as a researcher and an ethnographer in the four-year period I engaged with these labs. There were the long winding interviews (Annexure 7) – some conducted in the lab with Dharmadikari and his current students, one with Dharmadhikari's wife at their home and others with former students, collaborators and former colleagues in their respective homes or current work places. I managed to sit in for the PhD defence of one of Dharmadhikari's students in the university and took copious notes of what transpired. I attended Dharmadhikari's farewell seminar organised in the department and, when the opportunity allowed, other nanoscience- and technology-related seminars, workshops and conferences around the country.

I also spent a lot of time in the labs – one part of it chatting with the researchers as they went about their work, another rummaging through the racks and cupboards full of paper – this included Master's, MPhil and PhD theses; notebooks crammed with scribbled notes; one set of files full of scientific papers; another set with invoices, bills and receipts (Annexure 2) and a third that had correspondence of various kinds. Some of the best time I personally had was when I was photographing the labs – how does one meet the eternal challenge of visually capturing and conveying the essence of the space and the subject of the photograph? I photographed the lab spaces, the researchers as they worked, the instruments that were under construction and also those that already sat on the workbenches. A big sense of joint achievement was when one of these photographs (Image 5.4) made it to the cover of the May 2013 issue of *Current Science*, India's leading scientific journal. And then there was also the more relaxed, winding-down time, chatting, sipping chai and munching glucose biscuits in the canteen I had first met Dharmadhikari.

The account that you will read in the following pages is the outcome, via all the above, at documenting and structuring the journey of the laboratory from my window and from within the framework of Science and Technology Studies (STS) that was as new to my subjects as the STM was to me when I first heard about it. At the heart of this multi-dimensional exercise was an attempt to answer three main questions, each an independent one in itself but also inextricably linked to the other:

(a) What actually happened in Dharmadhikari's lab? What did Dharmadhikari actually do?
(b) What can/does this story tell us about scientific research and innovation in India? and
(c) How does one locate this work within the larger narratives and discourses of innovation in the Indian context, but also globally?

And all of this, of course, was subservient to my larger doctoral research project, which spanned six years and five nanoscience and technology labs in three cities of India. Dharmadhikari's labs were a crucial component of my journey and it is only befitting that this book should be about their story and their journey.

1.4 The structure of the book

This account of that journey will be woven around two broad pivots. The first are the details of the work done by Dharmadhikari, his students and their labs, more specifically, the instruments they made and worked with. The second pivot is related to the contextualising frames, the two most prominent of which are the political-historical context of the S&T enterprise in modern India on the one hand and the larger discussions on innovation that have increasingly come centrestage in relation to what scientists and researchers do, and what they *should* do.

In Chapter 2 that follows from here, I will first sketch out the larger story of the scanning tunneling microscope (STM), the instrument that was at the heart of Dharmadhikari's enterprise and, indeed, of this book. I will outline a brief history of the STM – from its creation in Western Europe, the Nobel prize of 1986, Dharmadhikari's enrollment into what has been called the 'instrumental community' and the story of the first STMs that he made in Pune. Chapter 3 will be a brief journey into and through modern India's S&T enterprise, right from the time of independence in 1947 to the turn of the century and beyond. It will detail the expectations that the modern

Indian state had and continues to have of its scientists and scientific institutions, which have to now also increasingly deal with the demands of the market and of a neo-liberal economics. These are the sometimes competing, sometimes complementary, challenges that Dharmadhikari and others of his generation have negotiated in ways that we have little information or understanding of. The fourth chapter travels through what might be considered an unexpected world in the context of innovation and of S&T in modern India. This is the world of jugaad, an idea of innovation and problem solving that is as quintessentially Indian as anything might be. There are many lines of discussion on jugaad and my effort will be to consolidate the available literature and simultaneously present an account of the wide-ranging, even wildly differing, arguments and positions different people have on it. In Chapter 5 I pick up the threads from Chapter 2 and present the empirical heart of my research and this book. I will detail the specifics of the instrument making and the science done in Dharmadhikari's labs over a period of two-and-a-half decades. I go into the nuts and bolts of what happened here, how these early instruments were made and what they were able to deliver. I will also discuss some of the key reactions and critiques and also outline the responses of the scientists themselves to the questions asked and the challenges put up before them. In the concluding section of this chapter, I offer what I consider the first of two main conceptual contributions of my work. This is the idea of technological jugaad that I characterise with the help of eight characters, one of the most important of which is that of reconfiguring materiality. In the following, Chapter 6, I pull out once again to present an overview picture that locates the details of the foregoing chapter and the specifics of this laboratory within the larger frame of innovation studies. I first outline German economist, Joseph Schumpeter's continued and enduring influence on the innovation discourse around the world and particularly in India. I then discuss two specific macro level contemporary innovation and S&T-related formulations in the Indian context: India's Science Technology and Innovation Policy (STIP), which was released in 2013 and India Technology Vision (TV) 2035, which was released in the year 2016. Mapping these formulations onto the realities of life and work in a laboratory shows considerable mismatch and slippage and it is important to understand the implications of this as also the challenges and opportunities this throws up for policy making. Chapter 7 draws upon the idea of technological jugaad presented in Chapter 5 and details the second of the two key conceptual contributions of this work. Here I compare and contrast technological jugaad with forms and cultures of innovation such as jua kali and bricolage seen in other geographies of the world and make a case for the de-centred as also the de-centring of the cultures of innovation. It points to the existence of many alternative realities and narratives of innovation that do not get the attention they deserve. They are, in fact, being drowned out by the dominant ones that remain focused on the metrics of publishing, patenting and commercial viability. Chapter 8 is a kind of epilogue where I re-enter Dharmadhikari's labs to re-look at the world and its interaction with the labs. The tone here is 'ruminatory' as I discuss both the past and the future of scientific endeavours such as his. What is the final fate of those instruments and the labs that made them? What does it say of the world and of the culture of innovation that they are part of and in which they were created? And finally, in the postscript, I briefly discuss the possible agenda for a continued research program. There are many things that are indeed possible and the three that I list out as most relevant and promising include the need for a deeper engagement with the general idea of jugaad. We clearly need more detailed case studies of processes and objects related to jugaad. Also needed is a deeper engagement with similar conceptualisations in other cultures and languages and through a study, for instance, of the etymology of the word and the history of its use. The other agenda is related to the lab. There is huge need and an opportunity in the Indian context of a greater engagement with the lab (any lab) via ethnographies at various levels and stages. India claim to have one of the largest S&T enterprises in the world yet studies in and understandings of the spaces where this S&T actually happens are conspicuous by their absence. Changing this scenario is crucial if we want evidence-based policy making and indeed for the policies themselves to be relevant and useful. A correction is needed to both the destination sought and also the journeys we will undertake to reach there. It is a correction that is as urgent as it would be exciting!

Notes

1 Retrieved November 10, 2010, from http://physics.unipune.ernet.in/~cvd/.
2 The discussion on the bicycle work had him visibly perked up.
3 Å is the symbol for angstrom, a unit of measurement, the length of which is 10^{-10} m (one ten-billionth of a metre).
4 The image can be seen online. Retrieved from www.almaden.ibm.com/vis/stm/atomo.html.

2 1986–2014

Making of the STM

2.1 Nobel for the STM

The first significant images from a scanning tunneling microscope (STM) anywhere had been reported in late 1981 but it wasn't until 1986 that the instrument (Image 2.1) and its makers, Heinrich Roehrer and Gerd Binning of the IBM Research Laboratory, Zurich, Switzerland, gained acceptance and popularity (Binnig & Roehrer, 1986). The ultimate stamp of recognition came, of course, when the duo were awarded one half of the 1986 Nobel Prize for Physics[1] for the successful development of the STM (Binnig & Roehrer, 1986; Mody, 2011).

The potential and significance of the instrument, which is universally credited with having spawned the now ever-expanding field of nanoscience and technology (NS&T)[2] was reflected in the press release issued by The Royal Swedish Academy of Sciences (TRSAS) when the award was presented in October of that year:

> The reason for their success was the exceptional precision of the mechanical design. One example of this is that disturbing vibrations from the environment were eliminated by building the microscope upon a heavy permanent magnet floating freely in a dish of superconducting lead (. . .). It is evident that this technique is one of exceptional promise, and that we have so far seen only the beginning of its development. Many research groups in different areas of science are now using the scanning tunneling microscope. The study of surfaces is an important part of physics, with particular applications in semiconductor physics and microelectronics.
>
> (TRSAS, 1986)

The year 1986 also saw the beginning of Dharmadhikari's STM journey, not because of the Nobel Prize as one might be tempted to believe, but something more complex and serendipitous. It was in 1986 that Dharmadhikari

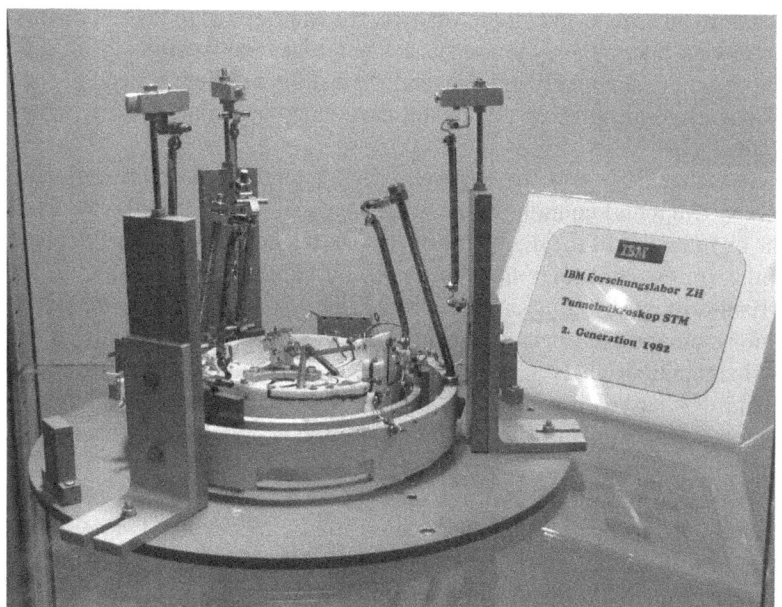

Image 2.1 A model of one of the earliest Ms built at IBM, Zurich
Source: Courtesy of CV Dharmadhikari

became part, unknowingly of course, of what Mody (2006, 2011) has called the 'Instrumental Community,' a community that was co-created over time and where probe microscopists were technologists and community builders at the same time:

> What initially brought probe microscopists together was a common interest in *instrumentation* – a new technology for peering at very small objects (. . .). Yet probe microscopists also set out to create a new *instrumentality* – a new way of doing things that would propagate beyond their laboratories and change the world.
>
> (Mody, 2011, p. 6)

It is significant here, particularly in light of Mody's conceptualisation, that Dharmadhikari's recruitment to the community was integral, but not central. Integral, because this was the community that became his primary point of reference; not central, because he does not seem to have had any significant influence on the community or its development and also because he remained on the geographical margins of the community that was located primarily in Europe and North America. His contribution to the community of STMers and to the STM itself is acknowledged only marginally even in the scientific community in India, and not at all in the more general history of contemporary (nano) science and technology in India, which is rather limited in any case.

Two elements stand out as particularly important in the constitution of this instrumental community as conceptualised by Mody (2011) – first, the significance of the discipline of surface science to STM development and, two, the core role played by a set individuals on either side of the Atlantic in constituting and sustaining that community. This included, among others, Heinrich Roehrer and Gerd Binnig of IBM, Zurich, in Switzerland; Nicolas Garcia of Universidad Autonoma de Madrid, Spain; Richard Colton of the Naval Research Laboratory (NRL) Washington, USA; and Paul Hansma from the University of California, Santa Barbara, USA.

These names are important not only for the STM community itself, but also for Dharmadhikari's journey and his recruitment as an STMer. He not only met all of these individuals in the initial, formative period when the STM and the STM community were taking off, he also happened to be present at the time and at the places where it was all unfolding – meetings, conferences, personal encounters and visits to key laboratories. Dharmadhikari was, most likely, the only Indian to attend the first STM conference that was organised by Nicolas Garcia in Santiago de Campostela, Spain, in July 1986 (Image 2.2), only a couple of months before Roehrer and Binnig were awarded the Nobel prize. The following year he travelled to the United States of America to attend the second STM conference, organised in Oxnard, California, by John Baldeschwieler. While in the USA, he visited among others, the IBM labs at San Jose and the Lawrence Berkeley National Laboratory in California and also met Richard Colton who was building his first STM while on sabbatical at the California Institute of Technology (Caltech). It was a relationship that built up over the years as they remained in touch and also visited each other's labs when the opportunity arose. In 2005, Colton was even requested by the Assistant Registrar of the University of Pune for an assessment and evaluation of Dharmadhikari's work. Colton not only responded positively, he even recommended that Dharmadhikari be granted professorship (Colton, 2005).

The significance of these travels that allowed for these meetings and exposure cannot be missed if one considers the context of an Indian scientist in the decade of the 1980s. Means of communication were seriously limited and travelling abroad was difficult and expensive if it was possible at all. Foreign exchange needed to travel abroad was a scarce and

Image 2.2 The first STM Meeting in Spain in 1986. Dharmadhikari (seated) is seventh from left in the first row

Source: Courtesy of CV Dharmadhikari

precious commodity and was strictly controlled by the government (Anderson, 2011). It was only the rich, the powerful or the fortunate who got this opportunity.

2.2 Surface science

One discipline that was central to the probe-microscopy enterprise and "particularly influential in the invention of the probe-microscopy community," was surface science (Mody, 2004, p. 27, 2011):

> Most famously, the 1986 Nobel Prize in Physics was awarded to the inventors of the STM largely on the basis of their contribution to the solving of an important question in semi-conductor surface-reconstruction research [and] by the 1970s [already], Bell labs and IBM [had] each employed more than a dozen surface scientists, in addition to numerous junior staff scientists and post – doctoral fellows.
>
> (Mody, 2011, pp. 16, 31)

The initial annual STM conferences were, in fact, sponsored by the American Vacuum Society, the professional society of surface scientists (Mody, 2004).

It is significant that this is the disciplinary background that Dharmadhikari belongs to as well. He got his PhD from the University of Pune in 1979 for the "Design, Development and Application of Field Emission Microscopic Techniques for the investigation of lanthanum hexaboride (LaB_6)/ tungsten emitters." He then worked, from 1980 to 1984, as a research associate at the James Franck Institute, University of Chicago, USA, with Robert Gomer, one of the best-known names in scanning field emission and ionisation. Dharmadhikari was awarded the 1980 Welch Foundation Award by the International Union for Vacuum Science Techniques & Application, and, in 1984, he even co-authored a paper with Gomer that was published in the journal *Surface Science* (Dharmadhikari & Gomer, 1984). Even today, after having spent a better part of three decades making STMs and working with these instruments, Dharmadhikari often refers to himself as a surface scientist.

In 1986 Dharmadhikari was in Berlin, in fact, to attend a field emission/ field ion workshop when he took what he calls "a detour" to attend the first STM conference in Spain and also "[visit] many laboratories (. . .), [see] many STMs [and the] (. . .) very hectic work going on in Europe. But," he added, "they [the STMs] were so complicated that I thought I could never make it in India" (Dharmadhikari, Interview, 02 March 2011).

2.3 The STM journey in Pune

In only a couple of years, however, most of this doubt seems to have gone and Dharmadhikari had established himself as an important player in the field. Enthused by what he had seen during his travels in Europe, he even sent IBM a proposal in 1987 to fund STM development and research in India. Nothing came of it, but it indicates the confidence he had in his own capabilities and the promise he perhaps saw for the field. When a joint Indo-US project titled "Scanning Tunneling Microscope (STM)" was initiated in 1988 with SS Wadhwa of the Central Scientific Instruments Organization (CSIO), Chandigarh, India, and Richard Colton from NRL, Washington, as the principal investigators, Dharmadhikari was drafted in as one of three Indian players who had experience and could make a significant contribution. In an email communication to me, Colton said that he recommended a greater involvement for Dharmadhikari and the Pune team, but for reasons not known to him, that never happened.

It was also in 1988 that Dharmadhikari successfully made and installed his first STM – this was under a staircase in the department as that was the only place available at that time. Students had also started working with him on the STM project and the first Master of Science (MSc) project related to the STM had been completed under his guidance in 1987 (Bendre, 1987). The first peer reviewed paper on an aspect of STM construction was published in 1988 (Bendre & Dharmadhikari, 1988) (see Annexure 3 for a facsimile of this 1988 paper); the first MPhil degree for the "development of a simple electronic control system" for an STM was awarded in 1990 (More, 1990) and the first PhD was awarded for STM-related work under Dharmadhikari's supervision in 1999 (Yehia, 1999). When I first visited the labs in December 2010, three students were in various stages of their doctoral work related to different aspects of scanning tunneling and atomic force microscopy.

In this period of nearly two and a half decades (1988–2014), Dharmadhikari and his research group had made a series of probe microscopes, including a photon emitting–scanning tunnelling microscope (peSTM) (Dey, 2010); 12 students had completed their Master of Philosophy (MPhil) degrees (see, for instance, Chaudhary, 2002; Dambe, 1995; Iyyer, 1994; S. M. Patil, 1994; Sawant, 1994) and about an equal number were awarded doctorates, (see Chaudhary, 2011; Dey, 2010; Iyyer, 2006; Yehia, 1999) – all while working on various aspects of making these instruments. Many of these students then worked as post-doctoral fellows in leading institutions around the world: Wayne State University and the University of California in the USA; Madrid Microelectronic Institute, Spanish National Research Council (IMM, CSIC), Madrid, Spain; Institute of Molecular Science, Okazaki, Japan; Institute for Nanostructured Materials-Consiglio Nazionale delle Ricerche (ISMN-CNR), Bologna, Italy and *Centre national de la recherche scientifique* (CNRS), Marseille, France. Some eventually moved on to permanent positions in prominent institutions in India, such as the Indian Institute for Science Education and Research (IISER) and the Defence Institute of Advanced Technology (DIAT), both of which are based in Pune.

The research group also published nearly 70 articles including many in the world's leading peer reviewed scientific journals such as *Applied Physics Letters* (Godbole, Sumant, Kshirsagar, & Dharmadhikari, 1997; Sastry, Kumar, Datar, Dharmadhikar, & Ganesh, 2001), *Langmuir* (Chaki, Singh, Dharmadhikari, & Vijayamohanan, 2004), *Surface and Interface Analysis* (Datar, Patil, Iyyer, & Dharmadhikari, 2004; S. Patil & Dharmadhikari, 2002) and *Advanced Materials* (Kumar et al., 2001).

2.4 History and geography, space and place

This seemed to me like a substantial body of work and a significant set of outputs. I was admittedly amazed, even awed, that it had all followed from the first STM that had already been made in the 1980s and by using methods, materials and processes that were not just along un-anticipated lines but also counter-intuitive at the same time. I have also seen much admiration in the cross section of different people in India that I have spoken to, particularly when I mention that the STM is an instrument that is credited with having spawned nanoscience and technology, and that the IBM scientists who first made it were awarded the Nobel prize for it in 1986. This admiration can be very easily located in the aspirations and ambitions of the India of today. In a country that is increasingly aspiring to be a super-power in science (and much else) it is a subject of much discussion, concern and even castigation that the only Nobel Prize in science for an Indian working in India is more than eight decades old (Mashelkar, 2011b; Narlikar, 2003).[3] It comes across as very commendable to this citizenry, then, that an Indian scientist was successful in making a sophisticated instrument at the frontiers of science in such quick time.

But there is more to this story than just locating Dharmadhikari's endeavours and the development of the STM in a historical perspective. The details of what was contingent at that time and place are just as important. David Livingstone (2003, p. 7) has shown convincingly that "space (. . .) is not (. . .) simply the stage on which the real action takes place. Rather, it is itself constitutive of systems of human interaction." The central relevance of place and local conditions to knowledge creation has, in fact, been one of the key themes of Science and Technology Studies (Felt, Fouche, Miller, & Smith-Doerr, 2017; Henke & Gieryn, 2008; Knorr Cetina, 1995; Livingstone, 2003; Shapin, 1988, 1995), and applied to Dharmadhikari's situation we see that the geography of his location becomes key to what he did or, indeed, could do. There were certain conditions of location and of a material, temporal and social nature that, in retrospect, came together in recruiting Dharmadhikari into this instrumental community and allowed him the success that he achieved. One might even argue that there would have been no STM if Dharmadhikari had not been in Pune, if he did not simultaneously have access to the key global centres of STM development in the early years or he had not been a surface scientist to begin with. These were perhaps necessary conditions, but it also worth asking whether they were sufficient by themselves.

It is important to note that Dharmadhikari's first encounter with the STM had left him with the feeling that it was too 'complicated' and 'expensive' for him to make the instrument in India. That he was able to in spite of this initial reaction is more than evident from the brief historical account sketched previously, and this sets up the stage for two sets of framings – one, the micro and the contingent; the other, the macro and the contextual – that are needed to understand what really happened.

The first (of the micro and the contingent) is related to the specifics of what this one set of labs and this group of scientists *actually* did. How, for instance, did Dharmadhikari overcome the factors of the 'complicated' and the 'expensive' that were his initial reactions? What materials, people and skill sets did he recruit for this purpose? How did he find what he needed and how did he use what he found? What in essence were the nuts and bolts of his science and his instrument making?

It is, however, to the latter framing (of the macro and the contextual) that I will first turn my attention to. And this, itself, needs to be split into two sub-framings that might appear at first hand to be considerably contrary if not entirely conflictual. One is the larger historical narrative of modern Indian S&T and the other, perhaps unexpectedly, is a discussion on jugaad, a quintessentially Indian entity that has more dimensions, meanings and relevance than we have perhaps realised, let alone acknowledged thus far.

But first to a short exploration of the history of modern S&T in independent India.

Notes

1 The other half of the prize was awarded to Professor Ernst Ruska for his fundamental work in electron optics, and for the design of the first electron microscope.
2 For a nuanced refutation of this claim from within the constructivist tradition, see Mody (2011).
3 The 1930 Nobel Prize for Physics was awarded to CV Raman for his work on the scattering of light that is now known as the Raman Effect.

3 S&T in modern India – a brief history

3.1 Self-reliance as a core agenda

Many people I spoke to, including Dharmadhikari himself, noted that there was a culture of making instruments in the physics department at the university in Pune from the very beginning. This was inspired quite explicitly by the larger ambition of building a (postcolonial) modern nation state that was both self-reliant and also scientifically and technologically advanced (cf. Abraham, 2000; Anderson, 2011; D. Raina, 2003). These are central components of what David Arnold (2013, p. 361) has called "Nehruvian Science" – a science which simultaneously created a "space for postcolonial ownership and subjectivity," "sought to contest Western presumptions of a monopoly over science," was structured "primarily in relation to India's national needs and Cold War ambitions" and was finally, "a program of delivery, committed to redressing" the basic needs of a country plagued with ill health, poverty, and many other such fundamental social and material problems.

Central to this, of course, has been modern (read Western) S&T, the introduction of which has a long and deep history in India. In *Organising for Science* Shiv Visvanathan (1985) has identified three broad phases in this history – the Great Surveys of the late 18th century, the establishment of universities in the presidency towns of Calcutta, Bombay and Madras and the eventual rise of the industrial research laboratory. One might add a fourth, post-1990s phase, wherein science and the scientific laboratory have had to negotiate a world of rapid economic globalisation at the same time as they've become key constituents of these very processes (Krishna, 2013). Dhruv Raina identifies, on broadly overlapping lines, the following four phases of how the history of S&T and has been characterised in India: "British and French Orientalists studies of the sciences in India; pre-independence nationalist studies; the phase of post-colonial reconstruction (. . .) and the post positivist phase" (D. Raina, 2003, p. 20). Put together this indicates that the three most relevant frames that nanotechnology and also the work of scientists like Dharmadhikari occupies in a contemporary context are, concurrently, the post-positivist, the postcolonial and one with globalisation at its centre.

Among the central concerns in this history (and one of particular salience for scientists like Dharmadhikari) has been the serious shortage of resources available for scientific research in a poor country like India (Balaram, 1999; Mashelkar, 2011b) This had led from the very beginning to an overarching national policy (and rhetoric) where scientists and technologists were exhorted from the highest political levels to achieve self-sufficiency and self-reliance. This is evident in the various policy documents and in the many statements politicians have made over the years – for instance – in Jawaharlal Nehru's inaugural address at the 34th Indian Science Congress held in Delhi in January 1947, a few months before independence:

I invite all of you who are present here, young men and old in the field of Science in India, to think in these terms of India's future and become crusaders for a rapid bettering of the 400 millions in India. (. . .) The first objective, it seems to me, from any point of view and more especially from the point of view of Science, is to help in the building of a free and self-reliant India. India today has made its mark in the world of Science, more especially in Theoretical Physics and some other departments also. We have done well when we have hardly tapped the talent in India. We have only scratched the barest surface of the Indian people, and yet have done tolerably well and now, when I think of what we can do, and will do no doubt, when we open the doors of opportunity to a large number of people in India, then the kind of picture I see rather overwhelms me. If we could tap, say even five per cent of the latent talent in India for scientific purposes, we would have a host of scientists in India.

(Nehru, 2003)

And then again by Prime Minister Indira Gandhi in 1982:

Self-reliance must be at the very heart of S&T planning. There can be no other strategy for a country of our size and endowments. (. . .) Considerations like security, time factor, performance guarantee and costs often compel us to buy advanced technology from the international market. But in the ultimate analysis, neither true defence nor true development can be bought or borrowed. We have to grow them ourselves.

(Prime Minister Indira Gandhi quoted in Menon, 1982, p. 1276)

This idea of self-reliance in S&T has been at the core of policy in India and it keeps coming back in different ways in spite of the changing political dispensations and the economic ideologies that have seen a significant shift worldwide since the time of Nehru and of India's independence. Importantly, it under-girded the work of many scientific endeavours of the generation that Dharmadhikari belongs to.

3.2 History of S&T narratives

One way to understand the development of S&T in India, particularly in light of the framings and methodologies I have used for my research, would be to understand how the history of S&T has been written in India. It is important to note that this history is significantly influenced by the different developments in S&T and has been dependent due to institutional configurations on the institution of science itself (Chakrabarti, 2004; also see Habib & Raina, 2007; D. Raina, 2003). This history of S&T, Raina also notes provocatively,

> hastend[ed] to serve as an apology for science in India, both past and present (. . .). The object called science emerges, therefore, as a positivist given, and scientific activity as a pilgrimage of truth, both feeding into and feeding from the Nehruvian vision of science and technology as the vehicle of development.
>
> (D. Raina, 2003, pp. 109, 116–117)

This, in Raina's analysis, is the reason why critical scholarship in the history or the sociology of science has failed to emerge in India. This alliance of 'science' with the 'state' has characterised much of India's post-independence history and has been marked, according to some scholars, by two significant watersheds. The first was in the 1980s with the negative experiences of 'development' and the resultant public movements that catalysed what Dhruv Raina (2003) has called the post-positivist phase. The second was seen in the early 1990s with the opening up of the Indian economy and the processes of globalisation. More has happened in more recent years, and these are indeed the multiple and complex legacies that contemporary S&T and by implication researchers like Dharmadhikari inherited when they began their research and science journeys in the 1980s (cf. Krishna, 2013).

The post–World War II period, which is also the beginning of the post-colonial[1] era in India, was the period when science came to occupy a central place in contemporary culture around the world and to be considered ideologically transcendent. Technoscience was enrolled as a central player in projects of postcolonial nation building and "large-scale investments in projects such as the building of big dams and industrial manufacturing units (...) became ubiquitous in many newly independent nations, with science and technology as key signifiers of social and economic progress" (Khandekar, Beumer, Mamidipudi, Sekhsaria, & Bijker, 2017, p. 671). It allowed politicians who invested in science (and technology) to be seen as promoting no other ideology but that of development and modernisation (D. Raina, 2003), and saw the emergence in India of "a strong politico-epistemological contract between the state and science catalyzed by a Bernalist-Nehruvian vision of science" (Varughese, 2014, p. 22). This was at the heart of what Amulya Reddy has referred to as the "Nehruvian dictum: more science and technology – more industrialization – less poverty" (A. Reddy, 2009, p. 9).

Led by Nehru, the country's first Prime Minister, the creation of S&T institutions and the use and advancement of S&T became central to the development and destiny discourse in India (N. Tyabji, 2011). 'Nehruvian Science' (Arnold, 2013) and its belief in the power and capacity of S&T for the progress of the country became one of the key features of the modern Indian state (GoI, 1958, 2003; MST, 2013; A. Reddy, 2009) and is illustrated quite explicitly, for instance, in the Scientific Policy Resolution of 1958 (GoI, 1958), independent India's first such policy articulation. It reflects, not unexpectedly, what the political leaders were themselves envisioning:

> The key to national prosperity, apart from the spirit of the people, lies, in the modern age, in the effective combination of three factors, technology, raw materials and capital, of which the first is perhaps the most important, since the creation and adoption of new scientific techniques can, in fact, make up for a deficiency in natural resources, and reduce the demands on capital. But technology can only grow out of the study of science and its application (....). Science has developed at an ever-increasing pace since the beginning of the century, so that the gap between the advanced and backward countries has widened more and more. It is only by adopting the most vigorous measures and by putting forward our utmost effort into the development of science that we can bridge the gap. It is an inherent obligation of a great country like India, with its traditions of scholarship and original thinking and its great cultural heritage, to participate fully in the march of science, which is probably mankind's greatest enterprise today.

Every Indian citizen has also been charged in the country's constitution with the responsibility of developing a "scientific temper" along with "humanism and the spirit of inquiry and reform."[2] This, in the hope that it will bring about some reconciliation in a country characterised simultaneously by unfathomable poverty and disparities on the one hand and spectacular diversity and richness on the other.

Starting in the 1970s and early 1980s, a series of public movements[3] highlighted precisely these contradictions and conflicts further by challenging the predominant idea of development, and asking important questions of the ideas of modernity and progress (D. Raina, 2003; Varughese, 2014; Visvanathan, 2001). Visvanathan (2001, p. 3684) notes emphatically, in fact, that it is these movements and not the "academe or science policy centres" that provided both, the "great critiques of science" on the one hand and "the most significant impetus to Science and Technology Studies in India," on the other.

While policy makers and a majority of the scientific community were (and still are) committed to the technocratic image of S&T in social transformation (GoI, 2003; IANS, 2016; Kalam & Rajan, 1998; MST, 2013), scientists associated with social movements started to voice their opposition to this positivist conception of science. There was an increased acknowledgement of the complexity of issues and this went together with the acceptance of "science's inability to account for the technical, socio-economic and cultural consequences of the problems at hand" (Varughese, 2014, p. 22; also see Nanda, 2003).

The period that followed "has been marked by a quest for new (. . .) historiographies for studying the history of scientific and technological knowledge," and also "by a moving away of emphasis from expert knowledge embodied in large scientific institutions to social movements contesting government legislation, wherein scientific or technological knowledge is implicated" (D. Raina, 2003, pp. 38, 43).

This is linked quite inextricably indeed with the narrative of the postcolonial state and of postcolonial S&T.

3.3 Postcolonial S&T in India

Postcolonial science (and by extension, technology) in the Indian context has evolved from and emerged at the intersection of many divergent, even opposing perspectives of the colony, of modernity, of science and technology and of the very idea of the nation state itself (Abraham, 2000, 2006). A key player in this has been the "state scientist – science workers whose larger goals and objectives were drawn from the imperatives of state needs" (Abraham, 2000, p. 169). These state scientists, while responding to the call of nation development, were also expected to deliver science of a quality at par with the best, but with resources that were only a fraction of what were available to their counterparts in the West. The science had to be distinctly Indian at the same time as it was international. Abraham notes that modernity and the postcolonial should not be seen merely as chronological categories, and while the modern was generally understood to be about the West, the aim was to still create a modernity that was distinctly Indian. He uses the trope of "landscape" to explain how the geographical localities and specificities of prominent "Big Science"[4] projects in India were mobilised to "transform the contemporary social world in order to produce a *future* that had naturalised the signs of development and modernity" (Abraham, 2000, p. 185, emphasis added). Recent times have also seen the emergence of a new contract between science, society and the state (Krishna, 2013; Mallick, 2014; D. Raina, 2014) and it is, now, no more about the state scientist, the agenda of national development or the quest for international scientific credibility alone. The current reality is messier and far more heterogeneous than it has ever been in the past.

There are a number of additional influences visible today in the narratives and discourses of science, technology and of the production of knowledge. Mallick (2014) points, for instance, to the redefinition of the role of science and the university in the context of academic capitalism, to rise of the entrepreneurial university and to the articulation of new categories of knowledge production such at the mode 2[5] (Gibbons et al., 1994; Nowotny, Scott, & Gibbons, 2001) and the triple helix models[6] (Leydesdorff, 2005; Leydesdorff & Etzkowitz, 1998). None amongst these developments is, perhaps, more prominent than that of globalisation – the increasingly un-hindered movement of ideas, capital and human resources across national boundaries that kicked-in, in the Indian context, with economic liberalisation in 1991.

This, in turn, has created a new set of demands of S&T – visible as they are in the increasing prominence of the metrics of publishing and patenting, in the expectations of commercial viability and commercialisation possibilities of research outputs, and in the discourses around innovation – all of this even as the agendas of societal relevance and contributing to development remain conspicuously foregrounded. It is a manifestation of what Krishna (2013, p. 10) has described as the "changing social contract between science and society (. . .) with the onset of [the] contemporary phase of globalization from the 1990s."

The future that Abraham's (2000, 2006) postcolonial science had sought to produce can be seen as occupied by S&T research in the present, albeit with many layers of complexity added on. It is not enough anymore, as we have seen already, for S&T to be a partner in development or for S&T to contribute to the building of a powerful nation state; the demands of commercialisation, a globalising science, a more demanding civil society and the changing social contract of science have also to be negotiated with.

It is asking for the simultaneous negotiation not just of different histories and expectations, but also multiple temporalities, locations and geographies. The challenges are serious, even unprecedented, but then so are the opportunities.

One finds many resonances of this postcolonial rubric in the scientist's constant negotiating, for instance, between the demands of the state, the market and of the universe of science itself; of the synergies being attempted between traditional forms of knowledge and a modern science and technology and indeed of the '(in)appropriate' and '(in)correct' ways of doing science in a contemporary world. It is reflected in the questions (see Section 5.3) that were asked about the methods used by Dharmadhikar and his lab, the relevance of the instruments they created and also of the science they delivered. It is a key feature, as we will see, in the narrative about Dharmadhikari as it is in his own narrative of the what, how and why of his own science and instrument building.

3.4 Contextualising the methodology

There are multiple strands in the above discussion that are important for contextualising what I present here: first, that the dominance of the positivist image of S&T has prevented the emergence of critical scholarship in the history and social studies of science in the Indian academy; second, that the social studies of science that did indeed emerge were from an engagement with politics and social movements that were often in opposition to the state and to the scientific establishment and third, a contemporary contextualising via the postcolonial framing brings in many elements such as the demands of the market and of a neo-liberal economics that have muddied the waters even further. Put together they provide a key insight into the complex nature of the debates and discourses in India around science, technology, innovation, education and development. The narrative has to be historically and socially contingent as is it germane to the constitution of the contemporary S&T landscape of which Dharmadhikari was/is an integral part.

All of this also has another important consequence and one that is immediately relevant for the methodologies I have used. This is related in the Indian context to a fixation for the textual reading of science, a consequent neglect of the history of techniques and technology and to the continued ideological representation of technology as applied science, or even a lower kind of knowledge (D. Raina, 2003). There is also, as Amit Prasad (2006a, p. 219) observes, "a surfeit of academic analyses of science as well as government policy documents on scientific research in India, but these provide little insight into how particular techno-scientific researches are conducted in India."

There are very few ethnographic studies in the Indian context of life and work within the laboratory,[7] of the intersections of the many worlds within science and technology or indeed of society's complex interfaces with this science and technology. The specific details of work done by and in laboratories – the nuts and bolts of what happens there – is missing because there has been a serious deficit in the efforts at entering the black boxes of science and technology. It is precisely this deficit that this account of the making of India's first STMs and AFMs seeks to address, even if only as a small initial step.

This is also a narrative at the same time of the 'enculturing of innovation' – of the how and what of these instruments, the materials, processes and people that were recruited in the process, the bridges that were built and new meanings that were generated. That success was achieved is accepted, but the processes and journeys by which this success was achieved are not fully known, leave alone understood. This is a journey with many unexpected twists and turns, but, before we go into those details, there is the other critical if counter-intuitive detour that we need to take – this time into the world of jugaad, a world I have already qualified earlier as one which is as quintessentially Indian as anything can really be.

Notes

1 I use 'postcolonial' in the temporal sense in this particular instance. For a discussion on the wider framing of the postcolonial, see Section 3.3.

2 Part IV-A, Article 51-A (h) in the 42nd amendment to the Indian Constitution, 1976.

3 Some of the more prominent among these include the Kerala Sastra Sahitya Parishad (Kerala Science Literary Movement) led people's science movement in the southern state of Kerala (Nanda, 2003; Zachariah & Sooryamoorthy, 1994); the movement opposing the dams in the valley of the Narmada river in Central India (Sangvai, 2002) and also those arising from the gas leak disaster at Union Carbide's plant in Bhopal (Fortun, 2001; Hazarika, 1987; Jasanoff, 1994).

4 The examples he uses include the construction in 1996 of the Giant Metre-Wave Radio Telescope, near the city of Pune in Western India; the neutrino detection experiments in a disused gold mine at Kolar in the southern Indian state of Karnataka and India's first underground nuclear test in 1974 at Pokhran in the state of Rajasthan.

5 In Mode 2 knowledge production contemporary research is increasingly carried out a) in the 'context of application,' that is, problems are formulated from the very beginning with a dialogue among large number of different actors and their perspectives; b) that there is an 'heterogeneity' of skills and expertise that is brought to bear upon the problem solving process; c) there is 'transdisciplinarity' which is premised, in particular, on the idea of the 'transgressiveness of knowledge' and there is a co-evolution in the sphere of knowledge production and of societal institutions; d) that 'reflexivity' and 'social accountability' are an integral part of the research endeavor; and e) there are novel quality control methods, including those that are from society and are socially relevant.

6 Triple helix refers in essence to the enmeshing of the agendas of the university, the state and industry in knowledge production.
7 Some of the more recent include, among others, Amrita Mishra's (2009) six-month study of a bioscience laboratory "located in the Southwest of India," Amit Prasad's (2006a, 2006b, 2014) study of magnetic resonance imaging (MRI) in India, Thomas' (2017, 2018) year-long study of religion, modernity and science at the Indian Institute of Science (IISc), Bengaluru, Renny Thomas and Robert Geraci's (2018) very interesting paper on the culturisation of *Ayudha Puja,* also at the IISc and an account of the nanotechnology-for-retinoblastoma research in LV Prasad Eye Institute, Hyderabad and Sankara Nethralaya, Chennai (Sekhsaria, 2017).

4 Jugaad and its many worlds/avatars

It would be relevant to mention here, at the very beginning, that almost no conversation on innovation in India, particularly in the upper half of the country, can happen without a reference to jugaad. It is a term that is as often maligned as it is used with pride and it tends, therefore, to effectively sidetrack discussions on invention, innovation and creativity. It was for precisely these reasons that I had made a conscious decision when I began my research project in 2010 to *not* engage with jugaad at the outset and see where the discussions and learnings around innovation take me. As it turned out, and not unexpectedly I might say in retrospect, there can be no escaping jugaad. One might want to give it a miss, but it will not allow itself to be overlooked. I tried and, to put it very simply, failed. And failed quickly. I may dare say that no research, discussion or deliberation on innovation in an Indian context can afford to ignore this jugaad, not even in the context of a modern scientific laboratory.

4.1 Understanding jugaad

Jugaad is a word in many Indian languages such as Hindi, Marathi, Gujarati, Punjabi, Oriya and Mythili that are spoken north of the Vindhyan Mountain range; it is also a term that does not appear to have an easy equivalent in English. It is not just an inextricable part of these local vocabularies, it is an integral part of the way life is lived and the world negotiated. It means different things to different people in different contexts: it is used to explain the process of reconfiguring materialities to overcome obstacles and find solutions; it could mean working the system to one's advantage; it is used as a synonym in certain contexts for gambling and corruption; and across large parts of western and northern India, there is a self-rigged vehicle, that also goes by the name of Jugaad; it is as much a noun as it is a verb – concept, process and product all rolled into one at the same time – an idea characterised by plasticity and an articulation that has a wide range of meanings and usages that revolve primarily around problem solving or solution finding.

It is not surprising then that jugaad comes up repeatedly in discussions related to innovation with as many translations and interpretations as there are authors – "creative improvisation" (Krishnan, 2010); "developing alternatives, improvisations and make dos" (Prahalad & Mashelkar, 2010, p. 3); "an arrangement or a work-around, which has to be used because of lack of resources" (Rangaswamy & Sambasivan, 2011, p. 558); and "creative adaptation" and adjustments (Cappelli, Singh, Singh, & Useem, 2011, p. 95). One of the most evocative renderings of what jugaad means in its multiple facets is provided by writer-diplomat-politician and prominent chronicler of contemporary India, Pavan Varma:[1]

> There is an Indian expression and, like others, is quite impossible to adequately translate: jugaad. People are encouraged to use some jugaad when faced with a blank wall, or a difficult situation. Jugaad is creative improvisation, a tool to somehow find a solution, ingenuity, a refusal to accept defeat, initiative, quick thinking, cunning, resolve and all of the above.
>
> (Varma, 2004, p. 72)

The diversity and the richness are evident in the different ways it is translated, interpreted and used.

One strain of discussion in the popular media has an evidently feel good and celebratory air about jugaad – what Philip, Irani, and Dourish (2012, p. 14) refer to as "pleasurable or strategic essentialism" – making a virtue out of a situation of necessity and compulsion. The popular news magazine *India Today*, for instance, notes in the editorial of a special issue on innovation that, "The best translation of that word [jugaad] is a combination of innovation and enterprise (. . .). *Jugaad* to Indians was both instinct and inspiration. The drive for a better way out, after all, is in India's bloodstream" (Purie, 2010, p. 1). The editorial of another edited volume of case studies on innovation by the business news weekly *Businessworld*, notes similarly that innovation for many Indians is "jugaad – the way of finding solutions to great problems using little more than ingenuity and a can-do spirit" (Datta, 2010, p. 4).

The celebratory slant notwithstanding, it is evident that jugaad in these publications is dealt with only in a perfunctory manner. The term appears only in the editorial notes in both the publications and that too only in passing. The *India Today* issue, for instance, profiles 20 innovators and innovations that range across diverse fields such as traditional pottery, modern medicine in the time of the swine flu epidemic and the development of a pedal power driven machine for washing clothes. Jugaad, however, does not find a mention anywhere in any of the detailed accounts of these innovations. The *Businessworld* compilation similarly discusses 23 case studies under the broad headings of 'Business/Process Innovation,' 'Social Innovations' and 'Organisational Innovations,' but has no mention at all of jugaad.

Another illustrative account of jugaad is seen in the *New York Times* article by journalist Anand Giridharadas (2010) and in the book, *Jugaad Innovation* (Radjou, Prabhu, & Ahuja, 2012) – both, strikingly celebratory of their respective ideas and interpretations of the term. Giridharadas offers the concept and particularly the automobile named jugaad (see next section for an account of the jugaad automobile) as the great innovative capacity of Indian genius and the alternative to the crises of the recent economic meltdown. Giridharadas claims that "jugaad is a way of life (. . .) that has anticipated important movements of the 21st century, from open-source technology to cultural fusion," and yet he appears to deny it agency when he positions the global economic order as the primary paradigm within which innovation or its variants are allowed to perform.

Radjou et al. too discuss jugaad from within a corporate and need-to-make-profit framework: "in today's consumer-driven economy we know that it's more important to *commercialise* technology" (Radjou et al., 2012, p. 11). They outline six principles of jugaad – a) seek opportunity in adversity, b) do more with less, c) think and act flexibly, d) keep it simple, e) include the margin and f) follow your heart – all made with an explicit aim of helping "Western firms innovate and grow in a highly volatile and competitive environment" (p. 20). They use the category and the concept of jugaad retrospectively and, in my opinion, much too widely and expansively than it should be. Theirs is the classical imposition of the analyst's category, and what the actors have to themselves say of their work and their innovation has no space here. My disagreements with Radjou et al. will also be starkly evident in my characterisation of technological jugaad (Section 4.3). While Radjou et al. ask for the "margins to be included" when they talk of jugaad innovation, my contention is that jugaad in an Indian context is located primarily in the margins – where then is the question of including the margins? Further, and as I have noted previously, they offer jugaad innovation to the Western entrepreneur as a solution in a competitive and volatile business environment, while my contention is that survival, and not commercialisation, is the primary intention of jugaad. The jugaad might eventually become a commercial success, but in the first instance it lies outside the broad framework of the market place. If survival in the competitive business environment is what Radjou et al.'s jugaad is about, one might see overlap and agreement, but our frames and worldviews remain diametrically opposite just like our points of entry into the discussion and, indeed, to the conclusions reached.

4.2 An embarrassment called jugaad?

On the opposite end of the feel-good spectrum of jugaad is the recent, wide-ranging and damning account by Thomas Birtchnell (2011) where he makes two key points – first, the underlying chauvinism in the increasingly wide usage and adoption of jugaad that he links to the projection of India's future hegemonic potential; and second, that "jugaad impacts on society in negative and undesirable ways (. . .) [and] (. . .) is wholly unsuitable both as a development tool and as a business asset" (ibid., p. 357).

Located on another axis, away from the celebration and the chauvinism, is a space where jugaad encounters much scepticism and even denial (Datta, 2010; Krishnan, 2010; Munshi, 2009; Prahalad & Mashelkar, 2010). The inside flap of Porus Munshi's book (Munshi, 2009) notes, for instance, that "India is known as a country *not* of innovation but of improvisation – or 'jugaad' as they say in Hindi (emphasis added)." In a recent issue of the business newspaper, *Mint*, the editorial argues that "it is time we moved from the glorification of jugaad to the celebration of true scientific innovation" ("The Missing Revolution," 2012). Prominent lawyer and politician, P Chidambaram, was quoted similarly when he was Finance Minister of India in 2012, as saying that

> Jugaad is not innovation. It is a very corrupt way of looking at innovation. We are quite adept at getting things done. But what we really need is to appreciate and reward genuine innovation.
>
> ("Jugaad in not innovation: PC," 2012)

Recent opinion pieces in Indian newspapers (Nazareth, 2017; Pinto, 2017; Thakur, 2013) continue to characterise jugaad on similar lines – the use is seen more in its negation with jugaad being what innovation is not and certainly should not be. At best it can be a stepping-stone to the 'real,' the genuine and a more systemic paradigm of innovation.

Rishikesha Krishnan (2010) argues in his book *From jugaad to Systematic Innovation* that the journey that needs to be made is clearly away from jugaad and towards 'systematic innovation.' He acknowledges that innovation results from a complex interplay of policies, institutions and incentive structures and, further, that innovation systems of countries are influenced by "historical, political, cultural, social and economic factors and philosophies" (ibid., p. 44). Yet he notes:

> industrial innovation abilities in India can't be strengthened without a more widespread belief in the scientific method (. . .) [that] underlies

research and all forms of systematic experimentation and exploration. Unfortunately, India remains stuck in a more unscientific paradigm of innovation, often labeled as jugaad.

(ibid., p. 170)

There is neither a conceptual nor an empirical engagement with jugaad in the entire manuscript and yet it is dismissed emphatically. This message comes through strongly in different ways from the cover of the book itself, and the typeface in the title provides the first indication. The word *jugaad* is made up using five different fonts for the six different letters; *systematic innovation*, on the other hand, is rendered in a more formalised, uniform font, all in capital letters. Jugaad, the implication seems to be, is too playful, random and unsystematic for it to be taken seriously; what we need, therefore, is systematic innovation. There are also two images – line drawings – that appear as the backdrop to the title on the cover. Under the word *jugaad*, in a faint, fading, grey rendering, is a sketch of one version of the jugaad automobile. Towards the bottom and occupying a prominent part of the cover, is the bold outline of the latest high-end, mass-produced personal car – an automobile that, one might add, is out of reach financially for the vast majority of India's population and which, in any case, would be of limited productive value even if it could be accessed. This then is the object of desire, the object (and kinds of objects) towards which India and Indian innovation should move – with the promise as it were, of a modern, modernised future.

Other authors such as Prahalad and Mashelkar (2010, p. 6) dismiss jugaad similarly because "the term (. . .) has the connotation of compromising on quality." They prefer to use the term 'Gandhian Innovation' for examples such as the development in 1991 of a super computer by an Indian firm at a cost of US$20 million and the more recent 'Tata Nano,'[2] the world's cheapest car available at a price of US$2000 (Chacko, Noronha, & Agrawal, 2010). It is an articulation that comes across as conflicting with Gandhi's own accounts of and vision for science, society and technology (C. S. Prasad, 2001) and the more recent articulation seen in the Gandhi-inspired 'Knowledge Swaraj: An Indian Manifesto on Science and Technology' where the effort has been to provide

an earthy fragrance that brings together the ordinary majority, and with an innovative spirit that breaks the vicious cycles that many sectors have been trapped in (. . .). Indian citizens are thus seen as active contributors in the knowledge society and not as mere recipients of science and technology.

(SET-DEV, 2009, p. 5)

While jugaad itself does not find mention in the manifesto, it is evident that it will get a far more enthusiastic welcome in this framework of innovation in and for science and technology.

The position that the various authors have taken is evident and yet there are two elements that, though unstated, stand out in common in almost all these narratives. First, there is little, if any, empirical engagement with the jugaad that is being discussed. Second, and this is of particular relevance in the Indian context, there is no discussion at all of jugaad in relation to research and development in the formal S&T system here. If jugaad is indeed inferior, unsystematic and a compromise on quality as noted by a number of authors, it is not a surprise that it has no place in discussions about the mainstream S&T system of the country; S&T research is believed, after all, to be the holy grail of innovation, creativity and progress.

Herein lies a significant paradox, the exposition of which is the key contribution of this particular case. Jugaad, as I found out and the subsequent narrative will illustrate, appears to be alive and kicking in the modern scientific laboratory and the scientific method, and there is no compromise on the quality of the output either. Importantly, I am not using jugaad here to describe and characterise only what I as an STSer saw and interpreted in the laboratory; it was a term and an idea that the analyst (myself) and the actor (Dharmadhikari, the principle scientist) came to accept together as a concept that could be used.

The lesson is a clear one – S&T cannot be separated from other aspects of the society and culture of its location. If jugaad exists in all domains of Indian life be it industry, social enterprise, business processes or rural innovation and adaptation, there is no reason why its imprint will not be found in S&T research and development as well, even if it is of a modern nanotechnology.

4.3 Technological jugaad that made the STM

The best-known jugaad product in India, without doubt, is an automobile found in parts of the north, the west and the east of the country. It is a vehicle that is created using a non-standardised manufacturing process, is not registered with the relevant authorities and does not exist, therefore, within any formal legal framework. The interesting thing is that every such vehicle differs from the other, and the only thing that binds them together is that each is fabricated locally and by assembling different parts that are generally procured from other scrapped vehicles – engines, tyres, wooden planks, steering wheels, seats and even agricultural water pumps. There is no restriction on what is used and it generally depends on what is available at 'that place' at 'that time,' leading also to names that are varied and different – Jugaad (Jolly, 2009) and Maruta (Purie, 2010) in parts of northern

India, Chakda in certain regions of western India (Varma, 2004, p. 73) and the Vano (Sengupta, 2014) in what is an interesting twist of and take on the Tata Nano (pers. communication, Asis De) in rural Bengal in the east of the country (Image 4.1).

The automobile so created is, generally, a locally crafted solution to an immediate problem such as a bottleneck in transporting agricultural produce to the nearest mandi (whole sale market for farm produce) or to transport people in a landscape of limited connectivity and mobility choices. These automobiles have, in fact, been banned by the Supreme Court of India because they were not registered, not insured and failed to meet certain quality parameters (Anand, 2012; "SC Bans Farmers' 'Jugaad,'" 2013).

Another well-documented though less prevalent form of jugaad is the use of an existing artefact for purposes completely different from what is was originally created for – "materials put to uses few could have imagined" (Philip et al., 2012, p. 13). The best-known example of this is again found in parts of north India where washing machines are used to prepare *lassi*, the popular local drink made from churning curd, sugar and water at high speeds.

Image 4.1 The Vano automobile in rural Bengal
Source: Author

Similar empirical examples have been presented in Kline and Pinch (1996), which is an account of the different uses the automobile was put to in North America in the early part of the 20th century, in Radjou et al. (2012), of an interesting recent case where Chinese farmers were seen using washing machines to clean their vegetables and in R. N. Reddy (2013), where he describes how, in recycling and re-using electronic waste in the city of Bengaluru in south India, the informal recyclers learn new skills, extend the life and meaning of thrown away products and also make money in the process.

There are many such examples of jugaad found all over India even though the word itself may be found, as mentioned earlier, only in north Indian vocabularies. Evidently, this jugaad is a locally crafted solution to an immediate problem. It is often a personalised survival tactic in situations of obvious resource constraints and/or denial (Rangaswamy & Sambasivan, 2011; Varma, 2004), and this is a characteristic that is fundamental and crucial to keep in mind.

Notes

1 I would like to thank Rishikesha Krishnan for drawing my attention to this work by Pavan Verma. It is striking to note, however, that Varma's exposition of jugaad appears not in the 'Technology' chapter in his book, but the chapter titled 'Wealth'. The 'Technology' chapter deals only with India's much discussed Information Technology (IT) sector.
2 The car that was hailed as innovative and a big success when it was launched is now acknowledged, even by its producers and promoters, as a complete failure in the market (cf. Madhavan, 2013; McLain, 2013). Production of the car was recently stopped by its manufacturer.

5 Dharmadhikari's microscopes and technological jugaad

5.1 Reconfigured materiality

There is one thing that stands out in most of these cases of localised and contingent improvisation and innovation, and it is indeed at the heart of what I am proposing as 'technological jugaad.' It is the element of reconfigured materiality that is implicated very centrally in the processes involved – in putting materials to uses not imagined initially, giving them fresh meaning and purpose and creating new worth and value. My key intention is to narrow down from what is otherwise a many-possibility and broad-spectrum interpretation of jugaad, to focus attention on the making of the instruments by Dharmadhikari and his research group. It is this concept of reconfigured materiality and technological jugaad that I saw operating prominently in this microscope-making enterprise of more than two decades, and the following three quotes from different though related sources illustrate how this was done in these laboratories:

> There was a huge magnet and I got a bobbin – a plastic bobbin from a tailor and we had a coil on that. That coil was put in a magnet and we hammered it with a wooden hammer. Then we looked at the resonance frequency. Simple technique (. . .). Now with (. . .) the latest vibration system we are getting the same resonance frequency after 20 years (. . .). Then we developed one [STM] in a fridge [(Image 5.1)]. I had a student from the Middle East. He said I am leaving and what do I do with my fridge? He gave it to me – we removed the compressor and it was a good acoustic shell (. . .). It's a totally new concept – it was used for nanotechnology.
>
> (Dharmadhikari, Presentation,[1] 11 March 2011)

> Like some of the piezos we used from (. . .) ink jet printers. The older models used piezos, so we took out those piezos and used those. Or the buzzer element. Jugaad is something like the spectrometer we used for the tunneling and photon microscopy – we got it from junk, repaired

Image 5.1 One of earliest STMs that was installed inside a refrigerator shell
Source: Author

it, improvised and used it – this is a jugaad (. . .). I used to go to juna bazaar [junk market] and find out how much is the resolution of stepper motors. [For] (. . .) the older, hard disc stepper motors we found out that [the] information is not available. To develop techniques to measure how many steps it goes, (. . .) I think is jugaad, because you find one technique, you use another one, (. . .) plug them together and once you do it, you have all the technology that they already invented – [but now] for something else.

(Dharmadhikari, Interview, 02 March 2011)

STM operated in air is susceptible to (. . .) acoustic pickups, so (. . .) [the first] STM (. . .) was encased in a plywood box [(Images 5.5 & 5.6)] lined with foam and glass wool pillows. The thickness of the plywood used is 20 mm. [In the 2nd STM] the vibration isolation was achieved by keeping the STM head on a steel round base plate (. . .) [and] the whole system is suspended with "bungy" cords [(Image 5.2)] using [a] tripod-stand [(Images 5.1 and 5.5)]. It was found that resonance

Image 5.2 Using a bungy cord inside the refrigerator to help with additional vibrational isolation

Source: Author

frequency of the system is ~ 2Hz. To protect the STM from acoustic noise, the total system is encased in a fridge-case (from which the compressor was removed), since the fridge case [(Image 5.1)] has [a] metal frame [and] shields the STM from high frequency noise. [The] body has glass wool insulation, which protects the STM from acoustic noise. It was found that the acoustic signal inside the fridge is less than 2dB.

(Iyyer, 2006, pp. 51–52)

Discarded refrigerators (Iyyer, 2006) (Image 5.1), stepper motors from junked computers, tubes from car tyres (Datar, 2004; S. V. Patil, 2002) (Image 5.3), bungy cords (S. V. Patil, 2002) (Image 5.2), viton rubber tubing (Dey, 2010), weights from the grocery shop, aluminium vessels generally used in the kitchen, a tuning fork from inside a wrist-watch condenser (Image 5.4) (Kolekar, 2013) and bobbins from sewing machines are a few examples of what went into the making of the first prototype and the other probe microscopes that followed. For Dharmadhikari, one has to innovate with whatever resources, knowledge and infrastructure one has in hand because no other option is available. The parallel with other examples such as the jugaad automobile and the use of the washing machine mentioned

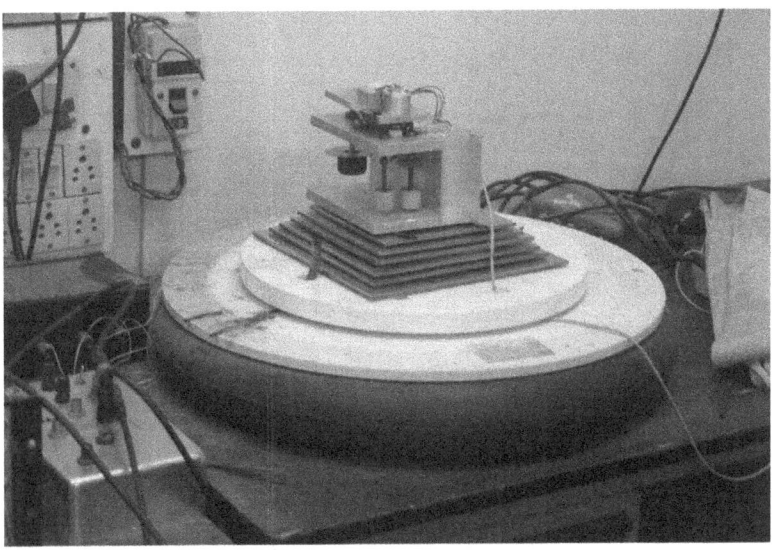

Image 5.3 A table top STM placed on the inflated tube of a car tyre for vibration isolation

Source: Author

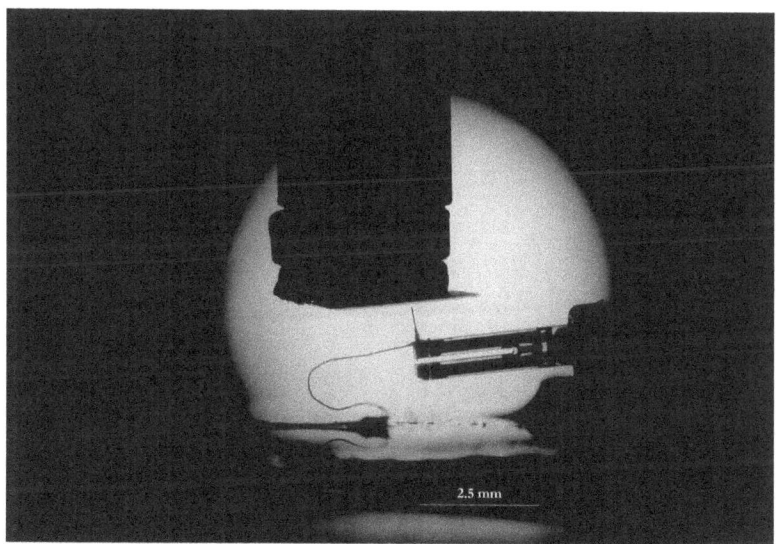

Image 5.4 The experimental set-up of the ultra-high vacuum STM-AFM[2]
Source: Author

earlier is immediately evident – existing materials and artefacts are used in completely new ways and/or are combined with each other to construct and operationalise a new idea or concept. It is not just reconfigured materials but also the thinking and processes and the context of location that have an important contribution to make.

5.2 Embedded in the local geography

Crucial to the success was a knowledge of the market – not just for the STM as a product, but also for the STMer as a consumer of the materials, skills and knowledge that were needed for making the STMs.

It was not good enough to only know what materials or skills were needed; one also needed to know where these could be found. A knowledge of the geography of the place and the relationships that built it was essential. Where would the old computers be found? Which small-time workshop would make the springs of just the right tension? Who would make the few components that had to be custom made? Where was the acoustics expert who would help design an ideal casing for the STM?

This attention to space and the relevance of the local to the process of knowledge creation has been a key theme in technology studies in general

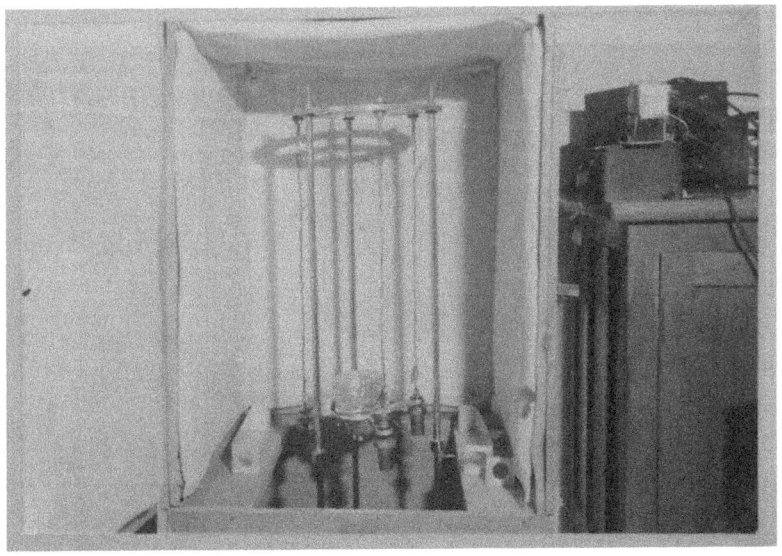

Image 5.5 A picture from the 2006 PhD thesis of SB Iyyer showing the STM set-up[3]
Source: Courtesy of SB Iyyer

Fig. II.2   Photograph of proper STM used.

Image 5.6 A picture from the 2006 PhD thesis of SB Iyyer
Source: Courtesy of SB Iyyer

and in lab ethnographies in particular (Felt et al., 2017; Henke & Gieryn, 2008; Knorr Cetina, 1995; Livingstone, 2003; Shapin, 1988, 1995). Livingstone (2003, p. 12) draws our attention to the "microgeography of the lab" (and also other similar spaces like the zoo, the botanical garden and the museum) to help understand that "matters of space are fundamentally involved at every stage in the acquisition of scientific knowledge." For Knorr Cetina, it is the power of the local which constitutes the power of the laboratory, and construction, therefore, is local too – "*with* local means and resources, with the equipment that stands around, the chemicals available, the technical skills and experience offered on the spot" (Knorr Cetina, 1995, p. 156). Shapin (1995, p. 306) has noted similarly that even though the "localist perspectives of science" have never been evaluated well, "science is undeniably made in specific sites, and it discernibly carries the marks of those sites of production." Henke and Gieryn (2008) take the example of mathematics in Göttingen in the early decades of the 20th century to explicate how in some instances "*places* ratify scientific claims" (p. 357). They argue further that even contemporary science, though increasingly globalised, can be seen as "emplaced science: [because] research happens at identifiable geographic locations amid special architectural and material circumstances, in places that acquire distinctive cultural meanings" (p. 353). This underscores the relevance of the 'context of discovery' even as it challenges the narrative that the local and the situated are not important in the case of science or of knowledge production in a more general context.

While an obvious resource constraint drove Dharmadhikari and his team's innovative and creative endeavours, the 'overall atmosphere of the city of Pune' was, as he emphasised a number of times, a very important player as well:

> There [was in Pune] a lot of awareness, support, software development, computers; engineering schools were coming up [and there were a] lot of R&D set ups. (. . .) We have [many] success stories in Pune (. . .) and [there were] these younger people who started their own companies with very modest budgets. [For] (. . .) any problem (. . .), right from image processing [to] development of hard [ware] (. . .) [they] would interact with us. My students will go and work in their factories and (. . .) it gave us a tremendous platform to work. That is Pune.
>
> (Dharmadhikari, Interview, 4 March 2011)

Not only were there a large number of educational institutions in Pune, but there was also a thriving small and medium industry enterprise located in the city and its suburbs where people were willing to try out new ideas, take risks and help others when something interesting was being attempted.

And there was also the other set of people – a small, but critical set of scientists in nearby institutions who were the users of Dharmadhikari's instruments, of the knowledge he and his research group produced and the different skills they had developed. It was a mutually beneficial relationship:

> A lot of people who were doing nano wanted to look at nano particles, DNA, and all kinds of combinations of nanostructures (. . .). Many times we did not get results and we ended up improving and developing newer techniques (. . .). My first interaction in nanotechnology was because of the people [who] were unable to get images on commercial instruments and so they approached me and I obliged them by some results or the other.
>
> (Dharmadhikari, Interview, 4 March 2011)

> Almost every laboratory in India would have approached us (. . .). Those who got good results thought their samples are very good [and] those whose results were bad said [there was a problem with our instrument]. (. . .) The important thing I noticed was they were coming in large numbers and it was the best way to test your instrument and improvise.
>
> (Dharmadhikari, Presentation, 11 March 2011)

Evidently, Dharmadhikari was well embedded in this elaborate geography and network of relationships and this was needed if he was to effectively capitalise upon and tap into it for achieving the ends he wanted to achieve.

The other important dimension that stands out for the diversity and richness of the skills recruited and the relationships used, was the incorporation of and collaboration with what one might call other ways of knowing or knowledge systems – the 'unexpected' players. This included, among others, a small-time entrepreneur with a spark erosion machine, a road-side workshop for aluminium sand-dye-casting and the procurement of springs from a workshop owner who would not understand the spring constant, but could deliver the needful based on Dharmadhikari's explanation of the requirements and the tacit knowledge embedded in his fingers. In another marginal case, Dharmadhikari even enrolled the traditional plating technique of kalai, the practitioners of which travel from house to house in rural and urban areas offering tin-plating services to housewives and restaurants for their copper and brass utensils.

Use of materials and skill sets that are available in and contingent to the context come across as key elements of Dharmadhikari's efforts and his success. It was not just part of knowing and being embedded in the geography and extended network of relationships, but also an acceptance and acknowledgement that valuable knowledge, skills and capacities lie in other spaces as well.

5.3 Critiques, questions, evaluations

Discussions with Dharmadhikari, with a set of scientists that either knew him or his work, and a perusal of literature in the context of Dharmadhikari's instrument building underlined many important critiques and questions. While some of these were of a general nature, others were specific to Dharmadhikari, his style of working and the instruments he and his students made.

These critiques and questions can be located within three specific domains, the first of which is related to the institution of the university. This includes broader issues of the state of India's universities, their challenges, capacities and outputs;[4] the imperative of teaching and the demands of research; the shortage of resources and the visible tension between the universities and the network of centrally funded research laboratories where mission-oriented science has grown faster in recent years than has academic science (D. Raina, 2003). Many of the university scientists I spoke to noted that these national research laboratories not only got a major chunk of the government's research funding, they also took away the cream of the students that were trained in the university departments. This was a vicious circle – good students didn't stay back because the incentives and infrastructure were not good enough and, at the same time, the absence of good researchers and research meant that the claim for more resources was a difficult one to make (cf. D. Raina, 2007; N. Tyabji, 2011).

The second domain of discussion revolved around the personal – on lines of what Lorenz-Meyer (2012, pp. 253–254) has conceptualised as the tension between the enterprising laboratory and the vocational laboratory:

> the binary opposites of East and West, local and global, static and dynamic that buttress (. . .) the superiority of the Western enterprising laboratory by equating it with outstanding international performance and excellence.

A similar binary was articulated by a senior scientist when he spoke of the "relaxed and non-ambitious way of Dharmadhikari's approach," where wisdom and knowledge mattered more than science and research, which was becoming all about publishing and moving aggressively up the career ladder. Dharmadhikari, he said, was the old-school professor who did pioneering work but was bogged down by his personality and the university system that neither gives incentives to scientists to excel nor pushes them to the limits of their potential and capabilities. In this, he felt, Dharmadhikari lost out and achieved much less than what he actually could have.

While these two domains are crucial to the context of Dharmadhikari's work, my main thrust will be on exploring the third domain of the discussions. This is about the instruments themselves and the issues that were raised around their construction, their operation and use, the results that they produced and the potential they did or did not have.

In March 2011, the university's physics department organised a national seminar on STMs in its premises to felicitate Dharmadhikari who had retired only a few months earlier. Those attending came from prestigious institutions from different parts of the country. All who spoke at the seminar – and this included senior scientists, collaborators, his students (Annexure 4), the incumbent and former heads of the physics department and vice-chancellors of the university – paid glowing tributes to Dharmadhikari, which is not surprising considering that this was a public forum. They were unanimous in their opinion that he had been innovative and pioneering. Many recollected and emphasised how his lab was the only destination in those early days for anyone who wanted to learn about or do something with scanning tunneling microscopy. In more private conversations, however, at least a couple of these speakers were (again, not surprisingly) more circumspect with their praise.

Nevertheless, these conversations and the other interviews I conducted pointed to a broad consensus that Dharmadhikari had indeed made a significant contribution to the development of nanoscience and nanotechnology in the country. Murali Sastry, one of India's more prominent nanoscientists is convinced, for instance, that Dharmadhikari's instruments were pivotal in the progress of his own research career and, for that matter, in the field in the country. Sastry was at the National Chemical Laboratory, Pune, when he first collaborated with Dharmadhikari and his STMs in the middle 1990s. The two of them jointly supervised doctoral students (Chaudhary, 2011) and, along with their respective students, collaborated over a long period of time, jointly publishing many research articles in leading journals (see, e.g., Dey, Pethkar, Adyanthaya, Sastry, & Dharmadhikari, 2008; Kumar et al., 2001; Sastry et al., 2001). The relationship and the benefits were mutual. Dharmadhikari acknowledged the challenges offered by Sastry in improving and improvising his instruments, while Sastry believes that he would have lost at least five years in his research work if Dharmadhikari's instruments were not here. It was crucial that those instruments, even with all their limitations, were available in those early years.

5.4 The commercialisation question

One of the most consistent critiques of Dharmadhikari's work was related neither to the quality of his instruments nor to the (jugaad) way he made them, but to the fact that he never tried to commercialise them. 'If these were such good instruments and they were producing good results why did he not take them to the market?' was a constant refrain. This was articulated most sharply by M1, a noted scientist and science administrator intimately familiar with the physics department at the university and with Dharmadhikari's work as well:

I feel that this entire approach and the development of an instrument could have been taken to a still higher order of proficiency and acceptability – [and to a] very clear and easily accessible instrument. Now, I think, this is where the group lost the lead (. . .). Your mind is always occupied with prototype instruments and prototype research – pure research – you see. So, I think that is where that convergence got lost. (. . .) [This] is a very qualitative judgement because I have not seen each and every research paper that has come. But whatever I have [seen] (. . .) gives me the feel[ing] that, yes there is an expertise, yes we had an early bird advantage, yes we created a prototype, we created many of them – every version a bit advanced version, but we could not convert that advantage into a production of such an instrument.

(M1, Interview, April 2011)

This concern and questioning is linked directly to the market paradigm and neo-liberal economic agenda (Krishna, 2013; Mashelkar, 2011b), which have made deep inroads into the laboratory and in S&T policy (see Sections 3.2 and 3.3 and Chapter 6). While there is nothing wrong with the framework itself, the question might be about its appropriateness in this particular context. What happens when the preoccupation with entrepreneurship, profit and a certain mode of economic and industrial production is sought to be imposed on a completely different endeavour without factoring in those contingent imperatives in any way whatsoever? Forcing the question of commerce and the market on a scientist or a researcher who has no claims to or interest in entrepreneurship is akin to asking a mass-produced product why it does not have a differentiated identity or an individualised aesthetic? The limitation lies in the frame of the innovation discourse, and the question, therefore, needs to be asked of this framework as well and not just of the scientist or the scientific enterprise.

Is it also not that Dharmadhikari and his students were not aware of or were not engaged with the commercialisation question themselves. Their experiences and responses were, in fact, very alive and sensitive to the idea of commercial viability – only their terms and frames of reference were different.

The practicality and viability of possible commercialisation initiatives was put into sharp perspective by Shivprasad Patil, who made an atomic force microscope as part of his doctoral work under Dharmadhikari's supervision and went on eventually to set up the Nano Mechanics Lab at the Indian Institute of Science Education and Research, Pune:

> In India it is rather difficult to (. . .) spin off companies like what they do in western countries (. . .). The reasons are many and not just because of the funding or because of our structure (. . .). If you look at mobile phones, there is a huge market (. . .), but suppose I want to make an AFM and sell it, there is no huge market for such things (. . .). You [also] have to have the support system in universities itself – incubation and other things and these things don't happen in India as much as they happen in the Western world.
>
> (Patil, Interview, 2011)

The reality on the ground is surely far more complex than it is made out to be. Dharmadhikari too acknowledged that the commercialisation idea was considered but his primary response was simpler and more direct as the following three short quotes, all from the same interview, illustrate effectively:

(a) We never thought of it [commercialisation]. We were so busy in publishing papers (. . .), doing science and I think the best way to test your instrument is to use it for (. . .) doing good science.

(b) At the same time I realised that doing your (. . .) own experimentation is always interesting – more interesting, more challenging. [There is] less throughput, of course (. . .), but in universities this is a better approach because you are training the students.

(c) I realised that if I can make simple [instruments] through the students, not only [will] we learn a lot about these techniques, (. . .) but we were also creating future generations which [was] proven later because most of the students got jobs in [the field of] nanoscience.

(Dharmadhikari, Interview, 04 March 2011)

The commercialisation question, interestingly, was being dealt with by changing the terms of the conversation – "doing science" and "training the students" were at least as important as thinking about the possibilities of commercialisation. For Dharmadhikari and his students, this was indeed the "vocational" and not the "enterprising" laboratory (Lorenz-Meyers, 2012).

5.5 A pedagogic tool

This training-of-students perspective also fits in well with the considerable discussion in literature on instrument making and doing things oneself as a pedagogic tool (D. Kaiser, 2005; Mody, 2004; Rasmussen, 1997), including in the specific case of probe microscopy in the USA (Mody, 2004, 2005). In a more general context, Mody and Kaiser (2008, p. 380) point out that at "the level of institutions, decisions over which equipment to build and which lines of research to support are also interwoven with decisions of which type of training to offer." Interestingly enough, and we see it in Dharmadhikari's case, the converse is also true – the kind of training capacity and expertise that is available with the institution and senior scientists will decide the kind of equipment that is built and the research that is performed. The (non)availability of certain financial and material resources meant that innovation in instrument making here would happen only along certain pathways – the jugaad pathway in this case. The value of this training and opportunity for hands-on work was most strongly articulated by Dharmadhikari's students themselves. This, for instance, is what Patil had to say:

> One thing that people blame you for is reinventing the wheel (. . .). I don't buy this argument at all because a lot of the people think that if I have a commercial AFM then I can work with it – why develop one? There are various reasons why you should build your own thing (. . .). If you have [an] AFM [from] the market, your ability of asking the question is itself limited because (. . .) [of] what a given instrument is able to do (. . .). No vendor, no one in the world, no company in the world, is going to customise for your needs and if they do they will charge you so heavily, you can't imagine (. . .). If, (. . .) right from your PhD, you are building your equipment, there is (. . .) this freedom, this kind of sublimeness [and] (. . .) it liberates you (. . .). The moment you buy one or two crores worth of equipment you are stuck (. . .) [and] your mind is tied (. . .) to the instrument. You do only what that instrument allows you to; [you] are scared to use it to its fullest capacity. (. . .) [What are] artifacts, what is true information, what is the false signal? Those things you know much better if you build your own stuff. People say you are reinventing the wheel [but] is not so, because it is really helpful at that time.

> (Patil, Interview, 2011)

The training, experience and confidence of doing things oneself held Patil in good stead between 2003 and 2007 when he was a post-doctoral fellow, first at Wayne State University in the USA and then at the Madrid Microelectronic Institute, Spanish National Research Council (IMM, CSIC) in Madrid, Spain (see Annexure 5 for a similar experience of another student). The benefits continue even today as Patil gets into more senior leadership positions himself and takes his training into his teaching as well. The other point he made is related to evaluation of the work in the lab:

> Look at (. . .) the amount of money that [Dharmadhikari's lab] has taken in and (. . .) [the output] in terms of publications [and] in terms of training manpower [it has delivered]. Publication is not the only thing that you should judge the research group by, [but also] training manpower. I think it is tremendous.
>
> (Patil, Interview, 2011)

Sastry (Interview, 14 March 2012) echoed the sentiment when he too noted that publications should not be the sole measure of one's scientific output. There was a lot of what he called "down time" in building an instrument, something for which there is never any acknowledgement! Naba Mondal (Mondal, 2015), Senior Professor at the Tata Institute of Fundamental Research and Director, Neutrino Observatory Project, has noted similarly in the specific context of high-energy physics that young researchers are not taking up instrument building because the pay-off time is long and they lose out in terms of publications to peers who are working in the lab or doing theoretical research. It is necessary, he says, that we create "metrics that recognise instrument development and retain these skills" (Mondal, 2015, p. 155).

This issue is also directly related to one of the main challenges that scientists and science administrators the world over are beginning to face and write about today. Two prominent examples of this are David W. Piston, Director of the Biophotonics Institute of the Vanderbilt University School of Medicine, USA, and P. Balaram of the Indian Institute of Science (IISc), Bengaluru and editor,[5] *Current Science*, one of India's most prominent science journals. In a short comment in *Nature*, Piston (2012, pp. 440–441) notes how

> over-reliance on automated tools is hurting science [and that] the research community must take more responsibility for teaching the coming generations (. . .) how to build, implement and troubleshoot their own experiments at the most basic level.

Echoing Piston's observations, Balaram (2012a, pp. 1241–1242) rues the absence in India of "trained technicians with a high level of competence in operating and maintaining facilities, [as a consequence of which] major facilities are sub-optimally used and sophisticated instruments are rarely exploited to their full potential." He notes further in a subsequent editorial that:

> many new institutions that are being created find it easier to acquire equipment than to recruit faculty (. . .). As Indian laboratories in

academic institutions accumulate the increasingly expensive tools and technologies of science, it may be important to remember that tools are only as good as the workmen who handle them.

(Balaram, 2012b, pp. 1383–1384)

This, one may argue, is precisely the kind of challenge that Dharmadhikari's way of doing science and training students is taking up successfully if one sees the capacities and the confidence of those who have trained with him.

This also raises the larger question of the state of instrument building itself in laboratories in India today. All I spoke to were of the opinion that this culture and these capabilities (and not just about probe microscopy), though prominently in existence, were always limited and restricted to a few research groups in any case. Even that culture is being lost now, fading away, thanks partly to issues around the increased availability of funding for research, for procuring (read import of) sophisticated instruments and to a reward system that does not recognise those who make these instruments – the "down time" is indeed not being acknowledged. It is symptomatic of a failure to recognise that "scientific instruments (. . .) and other material products (. . .) are constitutive of scientific knowledge in a manner different from theory, and not simply 'instrumental' to theory" (Baird, 2004b, p. 1).

The issue is very relevant in India today when the state is providing unprecedented resources for scientific research (DST, 2007; Manupriya, 2011; A. Reddy, 2009, p. 70). The case of the nanoscience and technology (NS&T) research is particularly illustrative of this changing scenario with huge financial allocations being seen through various arms of the Government of India (see Annexure 6 for a short note on the history and funding of nano-related research in India). Will the availability of these resources now change the way research in done in institutions and universities in India? Will those making instruments now stop making them because there is enough money to buy one off the shelf? Will the quality and quantity of scientific output improve significantly?

There are many questions that are beyond the scope of this present study but would be very interesting and worthwhile areas of investigation and research. Some of these are about jugaad itself, some are about science and the research that happens inside the laboratory, and others emerge when jugaad enters this laboratory and intersects with the science that is happening there. This does entail a larger and extended research project and I will come back, albeit briefly, to some of these (and other) questions towards the end of the book.

What remains to be done in this chapter then is to sketch out the key conceptual framework that emerges jointly from an understanding of Dharmadhikari's work, other secondary examples and the inferences that can be derived from them. I have called it technological jugaad and believe it offers as many possibilities of further empirical research as it does of conceptualising innovation not just in the specific and contingent but also in the larger general context.

5.6 Characterising technological jugaad

The account so far has presented the idea of jugaad in general and of Dharmadhikari's work in particular as a function of the milieu in which innovation and technological development is located. It is contingent on the influences, practices and ways of knowing which, as the STM example explicates, allows and even encourages the reconfiguring of material objects in varied though co-existing worlds. The junk market, then, becomes as important a resource for economic survival in rural India as it is for cutting edge science in the modern physics laboratory. This is at the heart of the idea of technological jugaad, which can then be looked upon as a 'culture of innovation.' Located within the larger frame of 'technological culture,' it is as much a manifestation as it is further evidence of the claim "that our modern high-tech society" can be better understood "by recognizing how its dominant cultural values and its technology shape each other" Bijker (1995a, 2006, p. 62). This is complementary to Harry Collins' (1985, 1987) "'enculturational model' of knowledge, in which knowledge is equated with culture and hence has an irretrievably social element" (Bijker, Hughes & Pinch, 1987, p. 308). It is also, at the same time, a convincing demonstration of how users matter within the specific social, political and cultural context (Oudshoorn & Pinch, 2008, 2003).

I will now use these detailed empirical and conceptual discussions and the extended inferences drawn from the same to characterise technological jugaad. This I do with the help of eight independent, but inter-related, characteristics, and I present them here mainly as signposts – pointers that can initiate further empirical research, help in gaining information and insights and also promote further discussion:

(a) *Reconfiguring materiality*: One of the cornerstones of the technological jugaad that we have seen, be it the automobile in rural north India or the STM in a modern physics lab, is the reconfiguration of materiality – giving new meaning to old objects (old refrigerators, discarded computers or automobiles, sewing machine bobbins) and finding uses that they were not initially created for.

(b) *Problem solving*: Technological jugaad mainly involves finding a solution to an immediate problem. The immediacy of the problem is often linked to economic survival, particularly in a context of resource scarcity. It may not be exactly the same in a physics laboratory, but can still be understood as an explicit manifestation of the imperative of continued existence. In that sense, then, it is different from invention or an activity of leisure such as pursuing a hobby.

(c) *Driven by resource constraints*: One of the key conditions driving technological jugaad is resource constraint and/or resource denial. There is, therefore, no option but to find new meanings and uses for existing objects – a direct linkage to the first characteristic of reconfiguring materiality just as it is linked to the last one – 'a culture of recycling.'

(d) *Bridging knowledge systems and ways of knowing:* This paradigm of innovation, as we have seen, particularly in the case of the physics lab, embodies an important element of interdisciplinarity. There is an awareness of what is happening elsewhere and, importantly, the capacity and the willingness to bring in ways of doing things that are located outside and, therefore, not considered part of the system.

(e) *Liminality*: In an earlier version of this conceptualisation, including in recently published work (Sekhsaria, 2013), I had articulated this characteristic as 'legally grey.' Following subsequent discussions with colleagues,[6] I realised that it is not about legality alone. Liminality would be a much better characterisation because the production process as well as the final objects created, like in the case of the STM and the jugaad automobile, lie across the borderlines of spaces such as law, regulations and standardisation.

(f) *Not (intended) for commercialisation*: Available evidence, albeit limited, suggests technological jugaad does not have much success in upscaling and commercialisation. Getting by (or survival) and not commercialisation is the primary intention of jugaad, though there is no reason why it should not become successful commercially. In the first instance, however, it lies outside the broad framework of the market place.

(g) *An activity of the commons*: The intellectual and material sources that are needed to find solutions in the jugaad way are not owned by anyone in particular and belong, as a corollary, to everyone. Technological jugaad can be seen, then, as drawing upon an idea of the 'commons' that is both conceptual and material, and where the resources exist all around – both in their usage but also, more importantly, in their accessibility and availability for use (Sekhsaria, 2011, p. 22).

(h) *A culture of recycling*: There is a way of looking at waste and a culture of recycling that is central to the jugaad enterprise. This is linked to conditions of a society where resources are scarce and access is limited. One has to make the best of what is available and this is also linked to formal and informal systems where waste, scrap and junk are indeed available for reuse in most towns and cities.

Evidently, many of these characteristics are integrally and inextricably linked to each other. Established categories get disrupted in the process because there is no longer a universal entity either of 'good quality' or of 'waste' that is only meant to be disposed. It is contextual, even cultural, as it helps re-engage with the idea of waste as "matter out of place" (Douglas, 1966) and of waste being a misplaced resource (Gregson & Crang, 2010; GTZ, 2000; Venkateswaran, 1994).

The inter-associations are constantly dynamic and interdependent, moving sometimes in parallel and sometimes in outright opposition even as they continue to intersect and weave their way through. Does a culture of recycling, for instance, allow for the reconfiguring of materials and of giving them new meaning? Is it a situation of serious resource constraint that forces a culture of recycling? Or, then, is it because people are embedded within multiple systems of doing and knowing anyway that they are able to find relevance and new uses for objects and systems discarded as waste or obsolete? A larger understanding lies in a combination of the answers, and studying more such examples and situations might reveal the existence of many different, though subtly tuned, permutations and options.

It would also be relevant to note here in the context of a (technological) culture of innovation that a large section of India's population lives in poverty and with seriously limited access to financial and material resources. Even more, a major chunk of the economic activity and employment is to be found in the informal sector (Kapila, 2010; Varma, 2004) where there is neither social security and nor any security related to employment or work (Kapila, 2010). It is this context of resource deprivation and/or denial that jugaad forms a bulwark to the livelihood and survival support system of millions. It is a value that is lost upon a number of critics who, as we have seen already, summarily dismiss jugaad on various counts. Technological jugaad might not perform precisely the same function inside a modern laboratory, but it is, undeniably, a part of the same continuum.

And because this is the case, it is my contention that there are many possible intersections with the policy discourse here that are bound to have significant implications for innovation and S&T policies, including for the processes of policy making.

Notes

1 The presentation was made at a 'National Workshop on Scanning Probe Microscopy: Techniques and Applications' organised in March 2011 at the University of Pune to felicitate Dharmadhikari on his retirement.
2 This was the latest in the series of instruments made by Dharmadhikari and his students that has a piezo scanner and a tuning fork (length: 2.4 mm) from inside a wrist-watch condenser.

3 This STM was assembled inside a padded up plywood box to isolate vibrations.
4 The primary body governing higher education in India is the University Grants Commission (UGC) (www.ugc.ac.in). There are roughly 600 universities in India that includes a combination of central and state universities (both state supported and private), deemed universities (also state and privately supported) and others like the Indian Institutes of Technology (IITs) that are called Institutions of National Importance (Banerjee, 2009; Deloitte, 2012; Ila, 2015). The S&T research infrastructure includes about 40 laboratories under the Council of Scientific and Industrial Research (CSIR) that are concentrated on applied research and a number of other laboratories concentrated on both basic and applied science that come under respective S&T departments of the Government of India such as the Department of Science and Technology (DST), the Department of Biotechnology (DBT), Department of Space (DoS) and the Department of Atomic Energy (DAE) (Banerjee, 2009; Ila, 2015).
5 He has since retired and also handed over editorship of the journal.
6 I would like to thank Samir Passi in particular.

6 Implications for innovation policy

6.1 What is innovation inside a laboratory?

The foregoing narratives – of the contexualising frames of the history of S&T in India, of the many meanings and understandings of jugaad and of the multi-layered empirics of Dharmadhikari's instrument building – provide us with key points of engagement with the current innovation policy discourse. An ethnography in the lab provides an insight into the real-world construction of facts and fact-constructing instruments in the laboratory; a look at historical developments contextualises the expectations from and the growth of S&T in a postcolonial Indian state; and a view through the window of a globalising world explains how the logics of quantitative evaluations, societal relevance and commercial viability have significantly infiltrated the thinking and the working in laboratories.

These were visible in all the labs, including Dharmadhikari's, that I researched for my doctoral dissertation.[1] One prominent drift of the discussions with researchers in these labs, for instance, was of the need to develop applications that are socially relevant. Many researchers articulated a position that it was wasteful to keep investing in basic science without the benefits ever reaching the large mass of this largely poor country. In many other conversations, commercialisation was highlighted as the primary measure of innovation and success, sometimes even eclipsing the novelty of the methods used or the value of the scientific knowledge gained in the research effort. Most discussions on innovation with these scientists veered quickly to issues of publications, patents, royalties and the success or failure in the attempts to commercialise.[2] At a conceptual level, it mirrors the "redefinition" of the societal role of science and of the university to now include concepts like the "entrepreneurial university and academic capitalism" (Mallick, 2014, p. 33). While considerable funding continues to be available for basic research in India[3] (nanotechnology is good case in point, see Annexure 6), the changing emphasis and expectations are visible. The standard and caricatured image of invention and discovery as the primary purpose of science in the laboratory has been supplanted with a lot more and in many complex ways.

6.2 Schumpeter's enduring legacy

There is no doubting, first and foremost, the continued and remarkable influence that the German economist, Joseph Schumpeter and his classic and much acclaimed works (Schumpeter, 1934 [2012], 1939, 1942) continue to have on the innovation discourse even today. It is not insignificant that, in a recent review of Innovation Studies literature (Fagerberg, Fosaas, & Sapprasert, 2012), two of Schumpeter's works have been listed as among the 20 most influential writings on innovation, and Schumpeter himself occupies the fourth position in a list of the 20 most important contributors on the subject. What stands out even more prominently is that Schumpeter's writings are the earliest in this listing – *The Theory of Economic Development* that was published first in German in 1912 (English translation in 1934) and *Capitalism, Socialism and Democracy*, which was published in 1942. The next piece of influential writing in this list does not appear until 1962, which is then followed by a cluster of the most recent in the 1980s and 1990s that created the influential idea of the 'National Systems of Innovation' (Freeman, 1987; Lundvall, 1992; Nelson, 1993). No one, it might be argued, wrote about innovation before Schumpeter did and, as has been noted by Fagerberg et al. (2012, p. 1135), "many ideas that are central in the innovation literature today can be already found in these works."

In his long introduction to a recent edition of Schumpeter's *The Theory of Economic Development* John Elliot notes that Schumpeter defined

> *innovation*[4] (. . .) as the commercial or industrial application of something new – a new product, process, or method of production; a new market or source of supply; a new form of commercial, business or financial organization.
>
> (Schumpeter, 1934 [2012], pp. xix, emphasis added)

This innovation, which is the "carrying out of new combinations" (ibid., p. xxi) is the linchpin of economic development. Central to Schumpeter's articulation of this process were the smaller business cycles that characterise capitalism and the larger waves of "creative destruction" that bring unprecedented change and turmoil with the introduction of new, radically different technologies (Kaplinsky, 2009, p. 18).

That much of the work in innovation studies continues to be focused on innovation in firms and industries (Fagerberg et al., 2012; Kaplinsky, 2009) is as much a comment on Schumpeter's influence as it is on the limited scope of research and theorising that has happened around the subject. As a consequence of this, perhaps, discussions around innovation became and continue to be circumscribed within a particular mode of capitalist production, circulation of goods, resources and values (see, for instance, Garcia & Calantone, 2002; B. L. R. Smith & Barfield, 1996).

At the core are technology, markets, entrepreneurship and profit and this is central to India's Science Technology and Innovation Policy (MST, 2013) (Section 6.3) and many of the newer conceptualisations of innovation such as Frugal Innovation (Bound & Thornton, 2012), Gandhian Innovation (Prahalad & Mashelkar, 2010), Reverse Innovation (Govindarajan & Trimble, 2012) and Jugaad Innovation (Radjou et al., 2012). The general term 'innovation' has come to be conflated with the more specific term 'technological innovation' on the one hand and with the imperative of making profit on the other. The issue of commercial viability and profit is particularly important because it has deeply infiltrated discussions in and around the scientific laboratory and the discourse of S&T policy in India. This, in fact, was the central core of the critiques and questions asked of Dharmadhikari and his labs (Section 5.4).

The imperatives of commercialisation and profit have become more central for the scientist and the laboratory today than they have ever been, and this thematic came up prominently and repeatedly in the articulations of scientists in all the different labs I worked for during my doctoral research. And yet, there was the implicit and explicit acknowledgement by these very scientists that the laboratory is not the 'firm,' the primary activity here is not economic and the key motive is not (and should not be) profit. When the work in the laboratory is over, when the results are out, when knowledge, the technology and/or the product exit the lab and enter a particular world of circulation and production – once all this has happened – one can argue that questions of profit, economics and financial sustainability will inevitably infiltrate that existence. But what about everything that happens till that stage is reached? What is innovation outside the realm of economics and outside the discourses of profit? What is innovation inside the lab? What are the frameworks one might use to understand this innovation? How, for instance, can one explain the work of the labs where Dharmadhikari and his students made their microscopes in contexts of resource scarcity by giving new meaning and a new lease of life to waste and the obsolete and by using ways of doing things that are located on the margins? Not to mention the fact that the science delivered was top notch and the scientists created were highly valued in that world.

Extended discussions, particularly when these entered more reflective territory, did reveal a clear ambivalence about the prominence of a commercial metric on the part of many of the scientists I spoke to. The undercurrent of a tension became visible in the scientist's urge to explore the frontiers of knowledge (the basic science), the simultaneous imperative of doing applied science and developing technologies that would be socially relevant and in the articulation of the logics of commercialisation and of profit. There was a new reality at this intersection and I could sense that the scientists and their institutions were grappling with its challenges and opportunities. This was their immediate reality and it was a response to the larger world within which it was located.

One way to understand this dynamic would be to assess how these laboratory- and institutional-level micro-experiences are reflected in the current-day meta narratives of innovation and of science and technology policy. How do the micro and the macro interact and interface with each other? What does one see as their relationship? Two policy formulations that are relevant here on account of being among the most recent are the Science Technology and Innovation Policy (STIP) – 2013 (MST, 2013) and India Technology Vision (TV) 2035 (TIFAC, 2015). Mapping their main thrusts onto the experiences of the labs and other spaces that they are supposed to be about suggests, as the subsequent section will illustrate, that these worlds don't square up. What one sees, in fact, is considerable slippage and a significant mismatch.

6.3 The Indian context – STIP 2013

The Science Technology and Innovation Policy (STIP), which was released at the 100th edition of the Indian Science Congress in Kolkata in January 2013, notes that:

> Scientific research utilises money to generate knowledge and, by providing solutions, innovation converts knowledge into wealth and/or value. Innovation thus implies S&T-based solutions that are successfully deployed in the economy or the society (. . .). Paradigms of innovation have become country and context specific (. . .). The national S&T enterprise must now embrace S&T led innovation as a driver for development.
>
> (MST, 2013, p. 2)

The short, 22-page document which was released by the then Prime Minister, Manmohan Singh, goes on to conclude with the following 'Policy Vision':

> The guiding vision of [the] aspiring Indian STI enterprise is to accelerate the pace of discovery and delivery of science-led solutions for

faster, sustainable and inclusive growth. *A strong and viable Science, Research and Innovation System for High Technology-led path for Indian (SRISHTI) is the goal of the new STI policy.*

(MST, 2013, p. 16, emphasis in the original)

The policy advocates that innovation must contribute to development in key sectors such as energy and environment, food and nutrition, water and sanitation, habitat, affordable health care and skill building and unemployment: "Science technology and innovation for the people is the new paradigm of the Indian STI system (. . .) [which] must, therefore, recognise the Indian society as its major stakeholder" (p. 2). There are also mentions of social good, inclusion, sustainability and of endogenous capabilities.

A closer reading reveals, however, that these concerns are perfunctory in the overall scheme of the policy and its primary thrusts. The edifice of innovation and of development is sought to be built on a foundation that is very much economic and technological, and this is clearly visible in some of the specific language used. The predominant focus throughout the document is on elements such as economic performance and growth, global competitiveness, high technology (p. 9), increased private sector investment in R&D and on technology development (p. 11). The role for 'society' is limited, primarily, to being the recipients and beneficiaries of the science, technology and innovation. The focus is not on how people can participate but on creating "delivery systems [for the] *diffusion* of scientific outputs and technology interventions into social systems" (p. 12, emphasis added), and for the systematic promotion across all sections of society of "the civilization aspect of science, or scientific temper" (p. 15).

A number of authors (D. Abrol, 2013; Mani, 2013; C. S. Prasad, 2014) have been critical of the policy, its focus and its thrust areas. This STI Policy, Dinesh Abrol (2013, p. 78) argues in his strong critique, "is quite high on rhetoric and intentions; it is rather weak on the issue of how the government will address the challenge of transformation of the systems of innovation in respect of social inclusion and sustainability." C. S. Prasad (2014, p. 55) notes that this policy "remains caught in a time warp (. . .), [and betrays] a weak understanding of how innovation is shaped in contemporary India and the world." He notes further that the policy privileges the technical expert as the carrier of knowledge and of scientific temper over the common citizen who is lacking both and is meant merely to be at the receiving end.

There is also the important question of how STIP took shape and what have been its primary influences? Not unexpectedly, the policy is in alignment with the dominant thinking of the S&T establishment on issues such as innovation, development, the perception of the public and the role of the corporate world in the current Indian context. See, for instance, the striking similarity in the perspective on innovation and in the language used in STIP and those in recent articulations of RA Mashelkar, one of India's most prominent and influential science administrators:[5] "A nation's ability to convert knowledge into wealth and social good through the process of innovation will determine its future" (Mashelkar, 2011b, p. 371). The following two quotes, the first one from his CSIR Founding Day Lecture in September 2011 and the second from an interview he gave me in June 2011, elaborate this idea further:

> Let's get into another fundamental. When we do research, we convert money into knowledge. The government keeps on giving us money and we keep on converting it into knowledge in the form of papers, patents, useful knowledge, not so useful knowledge, breakthroughs and so on. But actually it is innovation that converts knowledge into money.
>
> (Mashelkar, 2011a, p. 227)

> We have to realise that when we say (. . .) innovation converts knowledge into money, (. . .) all knowledge does not create money. It is only monetisable knowledge that can be monetised to create money. And that monetisability means that you must have ownership and that is where intellectual property came in.
>
> (Mashelkar, Interview, 08 June 2011)

The historical context is important as well. It is Mashelkar's 11-year tenure (1995–2006) as Director General of CSIR that is credited with having brought Indian scientific and industrial research in line with thinking in the West, with the 'market' (Mashelkar, 2011b) and in throwing up a new set of challenges and agendas for the science, technology and the industrial research establishment in the country (Pakrashi, 2003). He changed the "publish or perish" norm of the scientist to the eye-catching "patent, publish and prosper" (Mashelkar, 2011b, p. 58). The criterion of commercial and financial viability came, as a result, to be thrust upon the scientific community for the first time in India (D. Raina, 2007) – a good illustration of science's changing social contract with society after the 1990s (Krishna, 2013, p. 10). STIP remains truthful to, and even brings together, what Edward Constant (1987) has identified as the two traditions of technological change scholarship – one that believes scientific progress to be the key driver of technological change, the other that argues for the role of the market and of entrepreneurship to be considered dominant.

It is also not a coincidence that this attempt to change the image and expectations of the Indian scientist and the scientific establishment followed economic liberalisation in 1991 wherein the political establishment took the first significant steps in opening up and aligning India's socialist economy with the 'open' market.[6] It also marks a significant shift in the perceived role of the scientist – from being a partner in the Nehruvian vision of nation building and creating a scientific modernity to, now, also that of the scientist-entrepreneur who creates wealth and generates economic value from the science being done in the laboratory (Krishna, 2013; Mashelkar, 2011b). Mashelkar has, in fact, articulated the "freedom to compete" in an open economy after 1991 as the second of India's three moments of freedom[7] (Mashelkar, 2011b, p. 136). The continued influence of the Schumpeterian model is clear for all to see.

Importantly, a perusal of the details of what happens or has happened in laboratories like those of Dharmadhikari also illustrates how STIP is unable to account for the richness and diversity of the very labs it is supposed to be about and I will examine this further (in Section 6.4) by analysing recent discussions on the state of Indian science by contemporary research leaders ("Priorities for science in India," 2015).

It is particularly relevant in this context then to look at other articulations of innovation that are both multi-dimensional and nuanced at the same time. According to Gillian Marcelle, Executive Director, Research and Technology Park, University of the Virgin Islands, and a researcher with extensive innovation policy experience in a developing country context:

> Innovation is an intentional process of generating, acquiring and applying knowledge aimed at producing economic and/or social value. In developing countries, this process typically takes place through the unfolding over time of a wide variety of learning and capability building processes, rather than through the mastery of science and technological knowledge. Innovation is an investment effort in which, knowledge, financial capital, and other resources including cultural and social capital are deployed over time to create value. Deftly undertaken innovation can lead to the transformation of systems, values and culture as well as the production of new and/or improved products or processes.
>
> (Marcelle, 2017, p. 60)

Another specific formulation that has a particularly strong resonance in India is that of 'grassroots innovation' – an approach that allows and enables

> creative communities and individuals [including disadvantaged people living in developing countries] to develop alternative approaches (. . .) by converting their ideas into products and services [and] by blending modern science and technology, design, and risk capital.
>
> (Gupta, 2013a, p. 18, 2013b)

Embedded in the idea here is the explicit acknowledgement that "those at the bottom of the economic pyramid (. . .) are not, [by implication, also] at the bottom of the knowledge, ethical or innovation pyramids" (Gupta, 2013a, p. 18). The need and challenge is to build upon those resources in which "these poor people are rich" and to ensure that development is inclusive.

A look at these ideas of innovation makes STIP's limitations and narrow-focus evident and this is important because it is one of the most prominent recent policy documents of the S&T establishment in India. It is, on account of its authority and status, bound to play a key performative role in structuring the priorities and concerns of the scientists, in influencing formulation of subsequent policies and in making decisions for the allocation of resources and through these for the trends and trajectories it sets for future research. STIP has to be understood, it is my argument, as nested very much within the economics and profit-centred management and innovation-related literature from and about India that has come into prominence more recently.

Prima facie, STIP and other recent literature on innovation seeks to broaden the scope of discussion as is visible in the stated intention and in the language used. A close reading (Chapter 4; also Section 6.4), however, belies this promise and much the same can be said of another prominent overarching framework formulated recently by India's S&T establishment. This is the India Technology Vision (TV) 2035 (TIFAC, 2015) that was prepared by the Technology Information Forecasting and Assessment Council (TIFAC), an autonomous body under the Department of Science and Technology, and released by India's Prime Minister, Narendra Modi, in early 2016. A short engagement with this vision will help illustrate further the point that I seek to make.

6.4 India Technology Vision 2035[8]

India Technology Vision 2035 says it is an account of what we as a people and a country can be (and should be) in the year 2035 and claims to be rooted in the "collective aspirations of the people of India, the ambitions of our youth and the likely expectations of Indians in 2035 as the country grows" (TIFAC, 2015, p. 18). For TV 2035 "the technological 'peoples-cape' of India (. . .) [is] as important as its technological landscape. Fully cognizant that there is no India without Indians, TV 2035 speaks to – and of – all Indians" (TIFAC, 2015, p. 28).

And yet, and just like in the case of STIP, the details belie the expectations and the promises. In the first instance, and in spite of invoking complexity and diversity as the key constituents of an India of the present and the future, TV 2035 reduces Indians of 2035 to just six specifically articulated categories. Particularly relevant here is the section titled "Indians in

2035: Our needs" (pp. 36–46), at the heart of which is an illustrated sub-section that categorises the Indians of 2035 into six non-exclusive segments listed respectively as a) Rooted and Remote, b) Globalised and Diaspora, c) Left Out or Left Behind, d) Alternative Lifestyles and Worldviews, e) Creative, Innovative and Imaginative and f) Beehives and Production Lines. There are short accounts here of each of these categories, their possible percentages in the Indian population (the categories are non-exclusive so they add up to more than 100%) and details of each of their needs identified as something all humans are looking for: identity, prosperity and security (see Table 6.1).

There are many things to be underscored in the subtext of the language and the visual representations offered. For instance, the man representing the Rooted and Remote in TV 2035 wears a kurta and has a turban for headgear; the representative of the Globalised and Diaspora is a man of his youth in a suit and tie while the woman who represents the Left Out or Left Behind Indian is a dark woman (she is the darkest of the all the six humans illustrated), has long plaited hair and prominent rings hanging from her ear lobes. The question can of course be asked what is meant by being 'Left Out' or 'Remote' or 'Globalised' for that matter? Left Out of what? Remote from where, from whom? Why is it that the Beehives and Production Lines will be the largest segment of Indian society and why is it that they are ones who will be involved in "the productive process, that is the source of *all* resources and underlies *all* social existence"? (TIFAC, 2015, p. 40, emphasis added). What is meant by resources? What is meant by productive processes? Who decides what is productive? What are the parameters for these judgements?

Clearly, there is a normative assumption here of what prosperity (and much else) is, and one that has not been problematised at all. And, if this is a vision about India and about technology, what are the specific technologies that TV 2035 talks about? Indications are available in the small booklet titled *Technoscape* that accompanies the vision document and provides roadmaps for the 12 sectors[9] it has focused on specifically. The bulleted notes that constitute these preliminary roadmaps are deeply illustrative and I will use the roadmaps for Food and Agriculture (F&A) and Transport to further make my point.

The future of India's Food and Agriculture (F&A), if one goes by the *Technoscape*, will be exclusively technological, and that too of a certain type, suggesting as it does technologies that include advanced genomics and phenomics, robotic farming, hydrophonics/aquaphonics, nanotechnology applications, biofortification and apomyxis for fixing hybrid vigour and molecular manufacturing of food. It is most unlikely that the farmer, who in different ways dominates the political, economic and physical landscape of

Table 6.1 Categories of Indians in 2035

Category	%	Characteristics	Biggest need	Smallest need
Rooted and Remote	20	- Mostly rural - Adhering to old values - Seen as emblematic of good old days - Fear of becoming cultural exhibits - Though notion of remoteness will change because of road and internet access	Security	Prosperity and Identity
Globalised and Diaspoaric	30	- PIOs and their families - Considerable resources would have to be deployed to meet their interests - International comparisons and notion of 'best' practices would dominate - Extremely assertive about its rights - Concept of citizenship itself would change	Identity	Prosperity
Left out or left Behind	30	- Most deserving of technology policy attention - Society needs to acutely sensitive of inclusion issues - Won't get a chance or be able to keep pace	Prosperity	Identity
Alternative Lifestyles and Worldview	15	- Will opt out of the system - Follow alternative lifestyles - Different ideas about society and the good life - System should not interfere - System should remain engaged to gain benefits from mavericks	Identity	Prosperity
Creative, Innovative Imaginative	15	- Will be outside of straitjacket imposed by education, societal, governance systems - Much needed innovation will come from this section - Will be critical for the health of our country - Fount of economic and social dynamism	Security and Identity	Prosperity
Beehives and Production Lines	55	- Responsible for productive processes of country - Catering to them will be the biggest challenge	Security and Prosperity	Identity

Source: TIFAC (2015)

the country, has had any contribution to make in this vision and the visioning. If a majority stake-holder like a farmer is missing so prominently in the fine print, can one really expect that more marginalised sections that might include tribals, dalits, fisherfolk and industrial labour have been included?

Missing completely in this vision for agriculture, for instance, is the narrative of existing knowledge systems, attendant farming practices and traditional farming technologies that are visible even in the current mainstream scientific discourse in India for their qualities of sustainability, resilience and even productivity.(I. P. Abrol & Sangar, 2006; Maikhuri et al., 2015; R. Prasad, 2005; Ramani, Sharma, Monobrullah, & Mohanasundaram, 2015). TV 2035 has a vision for food and agriculture in which real-time agriculture and agriculturists are present only in name.

Future technologies for transportation, similarly, include among others fuel cell technologies, flying cars, magnetic levitation, flexible and foldable vehicles, JPod, hyper loop and high-speed pressure tubes. The citizen is missing here even more prominently. This vision of transportation appears to be a vision for vehicles, not for those who will be transported. The pedestrian and public transportation systems find only a passing mention in TV 2035. The concerns of the individual cyclist are missing, a discussion on non-motorised transportation is missing and the possibility that we can design our cities and towns differently to have an impact on transport, transportation and mobility is missing too. There is no reflection on the miserable state of transportation in our cities today, of the worsening air quality and how the systemic and systematic hostility to the pedestrian and the cyclist, the two most efficient and climate-friendly modes of transportation, increases with every passing day.

Citizens and the peoplescape that TV 2035 claims to include in the vision document are present primarily as recipients of the vision. The agency of the citizen and the possibility of doing things differently is conspicuous by its absence and is at the heart of the question that needs to be asked: can there be only one technology vision for a country like India? Should a country as big, complex and diverse as this one not have multiple visions? It is not a case of ought, but of is. This *is* a country and people of multiple identities, lives, lifestyles and aspirations. Multiple realities and multiple visions already exist. TV 2035 and STIP-2013 can at best be considered additions, albeit of a certain type, to a long list of possibilities and visions.

6.5 Policy implications

The foregoing discussion about STIP, about India TV 2035 and also the many narratives of innovation makes certain things clear. While the explicit claims in these formulations are of creating newer, more inclusive understandings of innovation and development, the details do not carry forward the agenda.

TV 2035, for instance, is a vision created almost exclusively by the techno-scientific bureaucracy of the country. Of the 24 names listed as key contributors, only a couple are from outside the formal S&T architecture of the country. The rest are all serving scientists or bureaucrats in institutions like the Defence Research and Development Organisation (DRDO), the Indian Institutes of Technology (IITs) and various laboratories of the Council of Scientific and Industrial Research (CSIR) or are former bureaucrats/administrators like former secretaries to the government or former heads of important S&T establishments. Citizens and the peoplescape that TV 2035 claims to include are conspicuous by their absence in both the vision and the process of visioning.

Jugaad Innovation (Radjou et al., 2012) and Reverse Innovation (Govindarajan & Trimble, 2012) too are explicit in their commitment to using ideas of innovation as solutions for Western economies and corporates deep in the throes of a serious economic meltdown. Radjou et al. (2012, p. 20) note that "adopting these principals [of jugaad innovation] could also help Western firms innovate and grow in a highly volatile, hypercompetitive environment" and become resilient organisations that think frugally, act flexibly and generate breakthrough growth (ibid., p. 228). Reverse Innovation, similarly, seeks to help multinational corporations innovate in emerging markets and to unlock opportunities across the globe (Govindarajan & Trimble, 2012). The metrics of evaluation are essentially economic and the thrust, clearly, is on growth, on profit-making and on expanding markets. The notion of the 'firm' and of economic activity as central to the innovation enterprise (Schumpeter, 1934 [2012], 1939, 1942) continues to be at the heart of the argument.

Part of the problem, Marcelle (2017, p. 73) notes pertinently, is that most of this theorising and conceptualising is done outside of the global south and also that these different conceptualisations end up being "labeled as categories of innovation, rather than (. . .) challeng[ing] the assumptions regarding the main dominant definitions of innovation."[10] This is the outcome and also the reason for an inadequate description of the specific conditions in which the innovation actually happens. The firm continues at the centre of innovation policy and theorising and the thinking endures, indeed, that it is not innovation if money is not made and if benefits are not accrued at the earliest. What is striking to note is the extent of which this thinking has now become a part of the scientists' frame and the constant challenge they face in negotiating with it.

This being the case, what is the view of scientists located within the mainstream establishment? What are their perspectives and concerns related to policies about the present of S&T in India and also its future? The thoughts and perceptions of some prominent contemporary researchers were put together recently in a special section by the journal *Nature* ("Priorities for science in India," 2015) and these provide very interesting insights. One prominent thing that stands out ironically about the monolith that delivers a STIP and a TV 2035 is the huge internal divergence and diversity. While KN Ganesh, director at the Indian Institute of Science Education and Research (IISER), Pune, identifies a deficit in funding and lack of world class infrastructure as key impediments in the development of science and technology in India (Ganesh, 2015), Vinod Singh, director, IISER, Bhopal, points to a failing tertiary education system (Singh, 2015) and Pradeep Mujumdar of the Indian Institute of Science (IISc), Bengaluru, to bureaucratic interference as the big problem (Mujumdar, 2015). Others highlight a range of other issues: Hiriyakkanavar Ila of the Bengaluru-based Jawaharlal Nehru Centre for Advanced Scientific Research points to the mismatch that exists between science and policy (Ila, 2015); Joyashree Roy, professor of economics at the Jadavpur University, Kolkata, points to an exclusive focus on the problems as a science-and-technology issue at the cost of attention to the socio-economic dimensions[11] (Roy, 2015); and Raghavendra Gadagkar, professor of ecology at the IISc, points to the constant attempt in Indian institutions to ape the "developed countries and the best endowed institutions in the world" (Gadagkar, 2015, p. 153).

Naba Mondal (2015), senior professor at the Tata Institute of Fundamental Research, Mumbai, asks for greater attention to be paid to those researchers who are interested in making instruments, while Gadagkar and Sunita Narain (2015), who heads the New Delhi–based non-governmental think-tank, the Centre for Science and Environment, ask, in different ways, for an increased attention in science to local problems like those of endemic communicable diseases, ground-water contamination, traditional methods of biodiversity conservation and sewage disposal. "Most importantly," Gadagkar (2015, p. 153) argues, "we should garland those who work on problems that are crucial to local contexts – even if they are of little interest to elite overseas universities or to 'high-impact' journals." For Narain (2015, p. 155), similarly, "The key obstacle is that everyday challenges are not top priorities for research and innovation. Indian science has always been fascinated by the 'masculine' agendas of space and genetics, not reinventing the toilet." Attention to the local and to the poor, they are arguing, will help develop S&T and an innovation and reward system that would be locally relevant.

A look at the institutions these researchers work with and/or represent also underlines a salient drift: eight of the ten institutions represented are an integral part of India's mainstream S&T establishment.[12] They are all located in metropolitan India (four in the city of Bengaluru alone) and are well-endowed, premier institutions that are able to attract the best talent and significant amounts of financial resources. They are part of the same S&T establishment that formulated STIP and yet we see that the primary thrusts of STIP don't map on to the comments, critiques and solutions that these researchers have articulated either. When STIP is unable to account for the diversity and messiness that exists within the formal establishment, one can only imagine the magnitude of the slippage if the different diversities, opinions, histories and interests that constitute Indian society are taken into account.

What these senior researchers in Indian S&T have to say in a prominent journal like *Nature* only complements the key point that I seek to make: the richness of methods, knowledge systems and worldviews that are an integral part of a laboratory whose life story we've just explored find very limited, if any, acknowledgement in the policy that applies to such a space. It might be difficult to believe that a story from the lab and STIP are, respectively, the micro and macro accounts of the very same space. The diversities that are made visible through ethnographies in the lab (and even in the opinions of prominent researchers) have been flattened out, even erased, in STIP.

Dharmadhikari's journey as a scientist making instruments and our own journey through his labs illustrate quite clearly the mismatch between the experiences in the laboratory and the key thrusts of macro policy formulations. Technological jugaad, and the scientist who might use such methods, has no place in STIP or in TV 2035. He might even be un-desirable: the scientist who uses waste, who uses unconventional methods, who uses other forms of knowledge, who uses what might be referred to as a jugaad way of doing things.

A journey through this lab (and the others that I studied for my doctoral dissertation) shows that life, work and innovation in laboratories in India is messy, multi-layered and multi-locational. It would be incumbent, therefore, on any policy formulation to account for and do justice to this reality. It is possible that policy makers are un-informed or ignorant of what happens inside laboratories, or, may be, they choose to ignore these realities. Either way, the policy implications are evident. The edifice of an effective and relevant innovation policy can only be built on a foundation that is alive to the realities of society and the S&T system it is meant to be about. It is unlikely that it will succeed otherwise.

Let me go one step even further from here to show that ideas of diversity and difference have already been operationalised in the context of innovation. It is my contention that different systems and cultures of knowledge and innovation are already in effective play, which leads simultaneously to a de-centring of the cultures of innovation at the same time as it makes a case for the existence, already, of de-centred cultures of innovation.

Notes

1 These labs were the Center for Advanced studies in Materials Science and Solid State Physics, Physics Department, University of Pune, Pune; Centre for Nanobioscience, Agharkar Research Institute, Pune; the Centre for Nanomaterials, International Advanced Research Centre for Powder Metallurgy and New Materials (ARCI), Hyderabad; The LV Prasad Eye Institute, Hyderabad and Sankara Nethralaya, Chennai.

2 This is one important part of the "cycles of credit" – the analytical construct deployed to understand and explain the motivations of a scientist. Other parts of the cycle include, among others, financial gain, positions and recognition that the scientists continually re-deploy and exchange with each other in their quest for growth and credibility (Latour & Woolgar, 1986)

3 It is relevant here that the public sector continues to play a dominant role in funding and supporting S&T in India. A 2006 survey revealed that 62% of support for research and development came from central S&T agencies that are directly under the union government (NSTMIS, 2006). Successive governments have, over the years, promised and also delivered substantive budgetary increases for the S&T establishment (cf. R. Raina, 2014)

4 For Schumpeter, innovation was "the carrying out of new combinations" broadly understood as the following: "(1) The introduction of a new good – that is one with consumers are not yet familiar – or of a new quality of a good. (2) The introduction of a new method of production, that one not yet tested by experience in the branch of manufacture concerned, which need by no means be founded upon a discovery scientifically new, and can also exist in a new way of handling a commodity commercially. (3) The opening of a new market, that is a market into which the particular branch of manufacture of the country in question has not previously entered, whether or not this market has existed before. (4) The conquest of a new source of supply of raw materials or half-manufactured goods, again irrespective of whether this source already exists or whether it has first to be created. (5) The carrying out of the new organization of any industry, like the creation of a monopoly position (for example through trustification) or the breaking up of a monopoly position" (Schumpeter, 1934 [2012], p. 66).

5 Mashelkar is a prominent scientist himself, is highly regarded and respected in political, corporate and S&T circles, has held a number of important positions in these sectors and is the recipient, most recently (in 2014), of the Padma Vibhushan, the country's second highest award for a civilian. He was Director General of India's Council of Scientific and Industrial Research (CSIR) from 1995 to 2006.

6 For a multi-sectoral overview of this process of liberalisation in India, see Balakrishnan (2011).

7 The first was India's freedom from the British in 1947. The third freedom in Mashelkar's articulation came in 2008 when India, led by Prime Minister Dr Manmohan Singh, signed Agreement 123, popularly known as the nuclear deal, with the United States of America. Mashelkar calls it the "technology

freedom" because it would allow India access to dual use technologies – those that could be used for military as well as for civilian purposes (Mashelkar, 2011b, p. 136; Interview, 08 June 2011).

8 I draw and quote extensively from Sekhsaria and Thayyil (2019, forthcoming) to make my points in this section. Also see Sekhsaria and Thayyil (2017).

9 The 12 sectors are Education, Medical Sciences and Health Care, Food and Agriculture, Water, Energy, Environment, Habitat, Transportation, Infrastructure, Manufacturing, Materials, Information and Communication Technologies.

10 Jugaad innovation as conceptualised and presented by Radjou et al. (2012) is an excellent example of this.

11 The specific sector that this observation refers to is the energy sector, but the same can be said of others too.

12 Of the other two, one researcher is at the University of Chicago in the USA, while the other is at the Centre for Science and Environment (CSE), a non-governmental organisation based in New Delhi.

7 De-centred/de-centring cultures of innovation

7.1 Culture/cultures of innovation

The idea of the de-centred/de-centring of the cultures of innovation is a response to the centrality and pre-eminence that certain ideas and notions of innovation have come to occupy in literature, policy frameworks and in the mainstream discourse. The diversity in the laboratory, the multiple realities of society and different realities of innovation notwithstanding, the focus of these dominant narratives is limited and centred on a few elements. S&T is one important locus, the corporate firm is another and the expert the third. Publications, patents and successful commercialisation are prominent metrics of evaluating success in the narratives of innovation, which continues to be seen as a linear process where the citizen is restricted to being a recipient. There is little place for recognising messiness, lack of control and diversity. The implication of this is that innovations are supposed to happen only in particular ways and in particular spaces – spaces that are endowed, for instance, with certain kinds of knowledge and access to power and resources. It is not stated in as many words, but it can be read as the subtext of formulations like STIP and TV 2035, which are as much about a centralised and homogenised understanding of a culture of innovation as they are about the centralising and homogenising of these cultures.

Central to this, of course, is the idea of 'culture' understood broadly as the entire way of life of a group or community of people as constructed by its activities, beliefs, customs, symbols, discourses, etc. (cf. Fischer, 2007; Geertz, 1973; P. Smith, 2001).[1] While this is an elegantly simple and useful explanation, a significant challenge arises in the Indian context on account of the concurrent existence of an innumerably large number of communities and their distinctly different 'ways of life.'

Culture in the Indian context, therefore, has a particular valence, and the complexity that arises on account of the great diversities of ecology, religion, caste, class, language, ethnicity, geography and history interfacing with each other and with the elements that constitute culture can only be imagined. Different sets of people live cheek-by-jowl and yet construct their lives and communities in drastically different ways. These lives also constantly intersect with each other in complex ways, making culture a subject fraught with tensions and challenges – it can be as much a unifying force as it can be divisive, and India's current cultural and physical landscape is a rich mélange of this reality. One important dimension of this society and culture has been syncretism, which allows for and creates space for the different cultures to live on simultaneously (Aiyar, 2015; L. Tyabji, 2015); these people, communities and cultures would not be able to survive otherwise. The implications for a culture of innovation are also quite clear then – people and communities that live life in different ways will seek to solve problems in different ways as well.

The key contention here is that the culture of innovation as articulated in a document like STIP is that there is only *one* culture of innovation just as TV 2035 speaks of only one vision for the future. Many other cultures of innovation exist simultaneously, and importantly these intersect with each other in different ways. Science and technology are important for sure, but there are other forms of knowledge and knowledge systems that also play important roles. The corporate sector surely innovates and produces wealth, but there are other constituents of society that also innovate and produce value of a different kind. While one culture of innovation moves in a linear fashion from research in the laboratory to a product in the landfill, there are other cultures of innovation that are inherently cyclic and iterative. Here, waste is an important resource (it is not waste at all, in fact), different experts are allowed to bring different kinds of knowledge to the table and the lay citizen also has understandings and insights that can be mobilised to find solutions. It is this simultaneous existence, relevance and interaction that I refer to when I speak of the de-centred cultures of innovation. And this applies, I argue, as much to different contexts and disciplinary fields as it does to different geographies.

7.2 Different cultures of innovation

To illustrate this and also take the discussion forward, I will pick up the idea of technological jugaad and use it to show that what is quintessentially Indian also allows for a more general understanding of the enculturing of innovation and of its de-centring. We have seen the conceptualisation and characterisation of technological jugaad as something explicitly Indian because of its location in a culture of languages across northern India and the broad sweep of its canvas from rural agricultural and poor urban India on the one hand to the modern scientific laboratory on the other, which considers the context of economic and resource constraints that circum-scribe life in large parts of the country, including in the laboratory – from a particular understanding of waste that allows for the reconfiguring of materiality to a contingent culture of recycling that feeds and is fed by the imperatives of survival and also to the existence of alternative, often tradi-tional, systems of knowledge.

In all of this, technological jugaad can be constructed as a stand-alone and a particular way of doing things that is confined to certain parts of India. What then of other places around the world with different cultural, social, economic and political conditions? Is it possible that technological jugaad with a different name, perhaps, may be found operating elsewhere too, albeit with localised variations and contextualised nuances?

7.2.1 *The reassembled cars of Taiwan*

The first 'other' culture of innovation I take up bears a striking resemblance to the jugaad automobile of north India. These are the 'reassembled cars' of Taiwan (Lin, 2009), automobiles whose advantages "include their safety, low prices, and flexibility in production and use, thanks to the collaborated network consisting of salvage yards and reassembled car makers" (ibid., p. 91). Local communities have different names for these cars such as 'iron cattle,' 'siqizai' or 'laqizai,' where:

> "Iron cattle" refer to reassembled cars that use single-cylinder motive power such as a water pump, a cultivator prow, or a motorcycle engine. "Siqizai" refers to reassembled cars that use four-cylinder engines; "laqizai" refers to six-cylinder engines.
>
> (ibid., footnote, p. 93)

Like in the case of jugaad in India, 'reassembled' in Taiwan has also had a negative social connotation of meaning unprofessional, unsafe or insecure. Lin also notes how the government along with the media worked together to put in place a strategy of stigmatising and regulating 'reassembled cars' to allow the auto-industry that was floundering in the 1950s and 1960s a foothold in the market. The negative image and repeated government clampdowns notwithstanding, the reassembled cars in Taiwan survived. In some cases, they were even being procured by government agencies like the police and local land offices and proved particularly useful in providing relief in disaster situations and also for transportation of farm produce.

In an insightful comment on safety and quality, Lin quotes figures to show that the accident rate for reassembled cars that were considered unsafe in the general perception was, in fact, much lower than that for mass-manufactured automobiles. The latter are built for a particular system of transport where the priority for high speed results in the use of thinner and lighter steel and where casualties are high when a crash does occur. These vehicles are similarly very unstable and unsafe when used in mountainous regions. The inference therefore is that quality parameters cannot be universalised and have to be looked at in the specific context they operate. Where are they being used, for what purpose and by whom? Exactly the same would apply in the case of jugaad and the same question needs to be asked of those who make a generalised statement that jugaad means poor quality. This is not to say that all jugaad is of great quality (whatever that might be), but to note that something cannot be dismissed as bad quality or unsafe just because it is jugaad.

7.2.2 Bricolage and user-driven innovations

Jugaad also finds resonance in the French idea of 'bricolage' (Levi-Strauss, 1962 [Translated 1966]), where the bricoleur "is (. . .) someone who works with his hands and uses devious means compared to that of the craftsman" (ibid., p. 16), "is adept at performing a large number of diverse tasks" (p. 17) and where the "bricoleur has no precise equivalent in English (. . .). [He] is a man who undertakes odd jobs and is a Jack of all trades" (p. 17, translators note in the footnote). Bricolage is made up of "elements [that] are collected or retained on the principle that 'they may always come handy'" and where none of the elements has just "one definite and determinate use" (p. 18).

There is an uncanny likeness between bricolage and jugaad and, in some cases, authors have even used the words interchangeably. Birtchnell (2011, p. 358) calls the jugaad vehicle, "a bricolage vehicle," while in the Hindi translation of the abstract of a paper on science education in a small village in rural India (Sharma, 2008), the translator who is a noted Hindi educationist, uses the word jugaad as a literal translation for bricolage in the original text.

The examples of technological jugaad that have been discussed also appear to have a prominent overlap with the now well-established idea of 'user-driven' innovation (von Hippel, 1988, 2005). In a study of the development in the West of scientific instruments across four instrument families – gas chromatograph, nuclear magnetic resonance spectrometer, ultraviolet spectrophotometer and the transmission electron microscope – von Hippel found that nearly 80% of this development had been done by users, the scientists themselves (von Hippel, 1988). The story of technological jugaad and of the development of the 'home made' STM in India then becomes even more interesting. The outcome (development of the instrument) may have been the same, but was the route followed in the West similar to the one followed in the Indian case? What were the kinds of materials used in the creation of the instruments? Where were they sourced from? At what cost, if any?

Valuable insights can be gleaned in this context from Cyrus Mody's (2006) engaging account of the commercialisation of scanning probe microscopy in the West. Mody notes the prominent role of commercial sources that early STM builders accessed to get components such as "vacuum chambers, piezoelectric crystals, [and] video output devices (. . .) geared to their specific applications" (pp. 62–63). He also notes that in other instances there was a "whimsicality (. . .) accompanied by *bricolage* in instrument building, [where STM] probes [were made] from pencil – leads [and] (. . .) AFM tips from hand-crushed, pawn-shop diamonds glued to tinfoil cantilevers with brushes made from their [researcher's] own eyebrow hairs" (p. 66, emphasis added).

While the three examples – the reassembled cars of Taiwan, bricolage and user-driven innovation – are drawn mainly from published literature, the one that follows is based on personal, first-hand experience from Africa. In December 2011, our PhD project group organised a workshop in Nairobi where we were joined by a small group of prominent nanoscience and technology researchers and S&T administrators from Kenya. Many of the scientists had returned after studying at prominent universities in the USA and in Europe and were struggling to set up facilities and the infrastructure to start research at home in Kenya. To this small group, I made a presentation of Dharmadhikari's successful STMs and of technological jugaad as a culture of innovation in India.

The trajectory of the discussion that followed was quite an instructive one. The first set of reactions were marked with serious questioning and doubts: how can such instruments provide nanoscale resolution? Are there instruments really realiable? How can they be standardised? If extended to other situations, will safety not be compromised? These doubts were discussed quite intensely for a while and then the tide, as it were, started to turn. Slowly the Kenyan scientists started to reflect on the situation in their context and in their backyards. Examples of innovation that they had heard of or seen in the field themselves were highlighted. It was argued that they had to look at what was already available with them and that they too had their own version of jugaad. This in Swahili was 'jua kali,' translated literally as 'hot sun' (UNEVOC, 1998), but used more expansively to mean 'hard work in the hot sun.' This was nothing but their version of culturally and socially embedded problem solving and innovation – 'if jugaad could work well in India, why wouldn't jua kali work in Kenya?' – particularly in a situation where there were serious resource constraints. Over lunch, one of the younger scientists even expressed his desire to visit Dharmadhikari's lab in India to see and learn exactly how these STMs had been made. A connection had clearly been made!

7.3 Finding a middle space for innovation cultures

Jugaad, siqizai, bricolage, user-driven innovation, jua kali: these are five terms from five different languages, cultures and histories that span the entire globe – yet there is something that ties them together. All appear interchangeably usable and at the same time there are factors embedded in the social, cultural and economic contexts that makes each one unique. Problems are solved, new ideas are generated and innovation happens in all these frameworks – in that sense one is like the other. At the same time however, they happen differently – de-centered both, in space and in action.

Jugaad and bricolage, for instance, are different in that they come from different languages, are based in different geographies and in the set of people of who use them. If use of "devious" means (Levi-Strauss, 1966) was, however, the parameter for evaluation, bricolage and jugaad (of which technological jugaad is a derivative) are considerably similar. The "whimsicality (. . .) accompanied by bricolage in instrument building" that Mody (2006, p. 66) talks of when describing a laboratory in the West, could well have been an account of what happened in Dharmadhikari's lab, only that the contexts were different. Observing that the jugaad-like elements used by Dharmadhikari in his STM building are ubiquitous in American universities, Cyrus Mody notes that perhaps there is

> an ideological vision of what counts as 'science' or 'high technology' that's shared by many in both India and the US but which gets applied asymmetrically – i.e., US scientists can put their STM in a refrigerator and be seen as creative bricoleurs, but when Indian scientists do the same thing they get criticised for relying too much on jugaad by the various anti-jugaad authors.
>
> (Pers. Comm., Email, 13 May 2013)

I see Mody's observation as an important comment on the nature of power in political and knowledge hierarchies at the same time as it is a call to correct that balance, to acknowledge that different cultures of innovation exist side by side and to accept that each is as valuable and productive as the other. There is no reason, for instance, that jugaad should not be treated with symmetry and with the same academic rigour and respect as one deals with bricolage or any other culture of innovation. It is to also acknowledge that jugaad is not an appendage to innovation like in the case of 'jugaad innovation' (Radjou et al., 2012), but a legitimate 'culture of innovation' in its own right.

User-driven innovation (von Hippel, 1988, 2005) too is a lot like technological jugaad, but with one key difference. Those doing the jugaad are not von Hippel's "lead users, ahead of the rest with respect to a related and important market trend" (von Hippel, 2005, p. 4). Whether it is the maker of the reconfigured automobile in rural north India or the STM inside the laboratory, those doing jugaad are more often than not trying only to 'catch up.' There is an overlap at the same time as there is a difference. Jugaad stands out as a survival strategy in situations of poverty and deprivation, and also for its (non)acceptance in the academic and other worlds. Bricolage and user-driven innovation are well-established and accepted conceptualisations; jugaad, on the other hand, evokes mixed feelings, and technological jugaad is not even part of the discussion yet.

What this implies is that different worldviews, knowledge systems and cultures of innovation are already at play, even if in a highly unequal world where many are forced to remain on the peripheries and have to struggle to survive. That many are not visible does not mean that they don't exist. Their invisibility is a function of the limited narratives and imagination of the so-called mainstream of science, technology, innovation, development and progress. Alternative narratives are possible – they already exist, in fact. All that is needed is for us to be sensitive to their possibilities and to be open to what they might have to offer.

Note

1 For a very interesting account of the relationship between religion and culture and, specifically, the culturisation of a Hindu festival inside the labs and workshops at the Indian Institute of Science, Bengaluru, see Thomas and Geraci (2018).

8 In the end . . . or call it an epilogue

A history of the STMs in Pune has been told, jugaad has been introduced and the enculturing of research practice that is visible inside the laboratory has helped conceptualise technological jugaad as a culture of innovation embedded in a social, cultural and historical reality. Having explored the lab, then having travelled into and through the world outside, I will re-enter Dharmadhikari's lab one last time for a final look. This one will be inside out and, not surprisingly, leads to a whole new set of questions and challenges on the one hand and raises afresh others that have remained barely hidden below the surface on the other. They are linked to the future of the objects that have been at the very core of this entire story – the instruments that Dharmadhikari and his students made and worked with for more than two decades.

8.1 A conversation with a student

The questions and the issues link up very intricately to the remark I made in the very beginning, where I noted that Dharmadhikari had retired just a few days prior to my first meeting with him and that I hadn't realised the full significance of it in that first interaction. The picture started to reveal itself about a year later in conversations with him and particularly with his student Sumati Patil (not related to Shivaprasad Patil mentioned earlier). Patil was the last doctoral student under Dharmadhikari's supervision in the university and had at least a year of research work pending during the course of my interactions with her.[1] These are conversations that deserve a detailed engagement because they revealed many aspects that lay barely hidden below the surface, but needed some coaxing and serendipity to reveal themselves.

It was late one evening that Sumati and myself chatted and shared cups of tea in the same canteen besides the badminton court where I had first met Dharmadhikari. It had been one of my longer days in her laboratory – I had first interviewed SB Iyyer who was one of Dharmadhikari's earliest doctoral students and then conducted a detailed interview with Patil. Part of that interview ended up being a monologue on what my doctoral research was all about. I told her about the nature of this research, the other work that I continued to do in parallel, what I saw as the potential of STS in India and what I had learnt from this experience. It was a very friendly and relaxed atmosphere and this contributed hugely to the conversation that flowed over the late evening tea. It is a conversation I would best call a 'ruminatory' one – full of questions, doubts and reflections on both sides – and one that made me realise the full import of the fact that I had entered the picture here at the cusp of a very important change.

I had arrived at a time that was not just a very crucial moment of Dharmadhikari's professional life (he had reached the age of retirement), it also marked the beginning of the end of the labs he had created many years earlier. The sense I got was that the STMs and AFMs Dharmadhikari and his students had built and Patil was working with would meet a very inglorious end. Nobody, Patil said, knew what would happen to the lab space and instruments in due course – probably a couple of years – when the lab finally 'closes down.' "People are just waiting to get the space," she said, illustrating her statement with the recent example of another professor. Instruments that he had fabricated or modified in his lab were all unceremoniously removed and left unattended in the department corridors once he retired and left. Financial and material resources, I realised suddenly, were not the only constraints that scientists and researchers have to contend with. Lab space – the real estate of the S&T world – was just as scarce and contested a commodity. There are many examples, and not just from this laboratory or university where labs and the instruments made therein have met such an end.

This reality had obviously been on Dharmadhikari's mind – there was no guarantee of what would happen to the instruments once he had moved on. This, for want of a better term, was an unexpected turn of events for me. I had had an enjoyable experience piecing together the story of these instruments, and the jigsaw was falling into place quite nicely and productively. It had been a fascinating journey to say the least and one that had thrown up numerous unexpected challenges and insights. All of this, I realised suddenly, was only part of the story. The curtains would soon be raised on act two of a drama that was still unfolding.

It was conjecture on my part then, and I imagined the situation playing out from an STSers point of view. An equally exciting chapter, many chapters perhaps, were still waiting to be written. Patil and myself hypothesised possibilities in many directions – these 'historic' pieces of equipment could move to some new place, or they could be taken out of the lab, be left to waste in the corridor, some godown or even be scrapped, *or* they could find a new lease of life under a new professor/researcher who would shepherd them into the future. We also discussed briefly the existence, possibility and the need of a space like a museum where equipment of this kind could be given a decent disposal/burial if I am allowed to use those terms in this context. Even if they are not actively used, they would/should at least be available for others, particularly the next generation, to see what happened in this lab and how things were done here. It had the potential of being a huge asset to the department and to the university to showcase its achievements and its history.

I suddenly realised the full import of Dharmadhikari's decision to allow me in the first time I met him and to then allow me access to his instruments, his students and his story. He had said this to me in the very beginning and again, in different variations, on a number of subsequent occasions. It was only while talking to Patil that evening, however, that I began to realise its full relevance. I had already been recruited by then, albeit unknowingly, as a player in the future of the labs. The object/s of study had turned the tables on the investigator from the very beginning; in fact, from the first moment of their engagement. If there was one reason that Dharmadhikari agreed to become the subject of my research without, perhaps, explicitly knowing it himself, this was it. He wanted the story told and hoped, maybe, that this telling will ensure something for the labs and the instruments that might otherwise be lost.

Why he did not think of doing it himself, leave alone making any effort in that direction, is a question I still don't have an answer to. Maybe it doesn't occur to scientists that it can/should be done, maybe they don't think it is important, maybe there is no recognition of its relevance, maybe they don't think it can be done at all. What is certain, for sure, is that the system does not provide any incentive and neither does it equip them with conceptual or methodological tools for such sociological and historical studies of S&T. When instrument making itself is not accounted for in the reward system, what hope might there be for the documenting and writing up of the history of their making? This is something that needs deep rethinking, even urgent correction; there will be no account otherwise of these stories and no account either of how many such stories are lost.

8.2 The story of another scientist

These realisations brought a completely new meaning to my research project – it was like the beginning of an afterlife. I had thought that in documenting and then retelling the story of these instruments, I was closing the loop on the story of the lab. Now, at the far end, I had stumbled on an opening that promised to balloon out big and wide. A change appeared to be waiting right around the corner that would be hugely different in its scale, but more importantly, degrees of magnitude 'different' in the nature, quality and impact of what could happen. It is about travelling and transformation, indeed about birth and death.

Before I go on to bring a closure to what I am narrating, there is a short detour I need to take to introduce the story of another scientist and another set of pioneering instruments. This is the story of Arup Raychaudhury, and while I am not yet in a position to present an account or understanding of its full contours, there is enough I know to re-emphasise the points I have already made.

I met Raychaudhury only once. This was on the sidelines of the International Conference on Nano Science and Technology (ICONSAT) held in Hyderabad in January 2012 where he had just been given the 'National Award for Research in Nanoscience and nanotechnology.'[2] At about the time Dharmadhikari was making his STMs in Pune, Raychaudhury was making his own in his lab at the Indian Institute of Science (IISc), in Bengaluru (then Bangalore). Like Dharmadhikari, Raychaudhury made a whole series of instruments, had a number of students complete their doctoral work making them and working with them and published scientific papers in the world's leading peer reviewed journals. That was perhaps the beginning of his journey into the world of nanoscience and nanotechnology, a field in which he is now considered one of India's best and for which his community had recognised him with a major national award. And here lies both the paradox and the huge tragedy. Raychaudhuri had just moved from Bengaluru to Kolkata as Director of the SN Bose National Centre for the Basic Sciences when we met, and he told us that his instruments in Bengaluru had all been junked – there was no need and no space for them now that he had left. Sumati Patil was with me during that conversation with Raychaudhuri and we could barely conceal our disbelief. Here is a short snatch from the recording I made of my interview with Raychaudhury that day:

(Raychaudhury (AKR)) They changed the department recently. So all my old equipment – they have junked it.

(Sumati Patil) Junked?

(Me) Junked it? They junked it?

(AKR) Everything.

(Me) Is it there? Lying somewhere?

(AKR) No. I don't know. I don't care where they are. I have one broken piece lying in Calcutta – you can [come and see it].

8.3 In the very end

Would this be the fate of Dharmadhikari's labs and his instruments? What was conjecture and speculation in the discussions I had had with Patil started to materialise in a couple of years. The stage, in fact, had changed completely by about the time I was finishing my doctoral research. Dharmadhikari's last two students in the university (Patil included) had finished their doctoral research and had moved on. More importantly, the movement of the instruments and the space they had occupied had also started. When I visited the physics department in the university in July 2013, one of the two lab spaces had already been withdrawn. There was no STM lab anymore (Image 8.1) and everything that made up that lab had been moved to the adjoining scanning force microscopy lab.

During a telephone conversation a few months later, Dharmadhikari told me with a sense of agitation and also of resignation that it was all over – that the second lab too had been pulled apart, that the instruments and the materials had been dumped somewhere, and the space had been vacated for someone else to move in. Twenty-five years of an intense scientific enterprise had been brought to a quick and ignominious end.

What insight, I wonder as I come to the very end of the story I have wanted to tell, does this final episode of a laboratory's journey offer us?

Image 8.1 The dismantling of the STM laboratory – a July 2013 picture
Source: Author

It clearly opens up a number of new questions, just as it simultaneously answers and closes down existing ones. I am not going into those details here, but there can be no doubt that a detailed exploration of this (really) final chapter of laboratory life will provide us important understandings into the culture of innovation that not only allowed for the instruments to be created in a particular way, but one that allowed the instruments to be created at all in the first place. It is the same culture, nonetheless, that did not have any space, even in its memory, for those very objects.

Notes

1 Sumati Patil finished her doctoral dissertation in 2014.
2 Attempts to meet him for another interview and a more detailed account and understanding of his story weren't successful because of time, resource and logistical constraints.

Postscript

A research agenda for the future

What I have attempted and presented in this book should, in my opinion, only be considered the tip of what is potentially a hugely challenging and exciting research program. Much more can be done and it is here that future paths beckon. There are three that I will outline in brief and many more can surely be added to the list.

An engagement with the lab

First and foremost, this calls for a lot more empirical work in the laboratory. There are as I have mentioned already very few ethnographies of work and life in Indian laboratories even though we claim of have one of the largest S&T enterprises in the world. We know very little of what our scientists do, how they do things, how they construct their labs, their instruments and their science. This is a vast unexplored terrain, an engagement with which will be hugely insightful and productive. There are many questions waiting to be answered about the nature of scientific practice, about the nature of innovation inside the lab, about science, technology and instrumentation itself, and indeed about the nature of jugaad in these spaces.

Illustrative and relevant in this context would be the email I received in May 2013 from Prof KL Chopra, former head of the physics department at the Indian Institute of Technology Delhi (IITD), New Delhi, and also former director, Indian Institute of Technology Kharagpur (IITK). This is what he had to say in response to an article of mine (Sekhsaria, 2013) that had just been published on Dharmadhikari's instrument making enterprise in *Current Science*, India's leading science journal (also see Chopra (2016)):

> I have read with much interest and appreciation your beautifully written article (. . .) in the May 10 [issue of] *Current Science*.

Dr Dharamadhikari, who I have met and have admired, deserves such a tribute (. . .).

When I joined IITD (. . .) as Senior Prof of Physics in 1970, (. . .) I was given two absolutely empty rooms to establish a Thin Film Laboratory. And, very little money (. . .). Having heard of my scientific contributions, several PhD students joined me immediately. We sat down together to clean the lab, and start building basic components and instruments such as vacuum chamber, vacuum systems, thickness monitiors, e-beam, plasma and magnetron systems, DLTS system, I-V and C-V plotters, etc etc etc. – all from junk material available from Chandni Chowk. My students used our own old lathe, glass blowing, welding set-ups, etc for fabricating equipment. Since our interest was to publish high quality research papers in the field (. . .) we did not record all our JUGAAD achievements. But, we did publish a lot and earned global recognition in no time.

This raises interesting questions just as it offers other possibilities. Does the jugaad we have seen in the STM example (and what Chopra articulates of his own work) exist elsewhere in mainstream S&T research too, or are these isolated cases? Where does jugaad stop and science start, or is it a wrong question to begin with? Is there a difference between the jugaad in the street and what we saw in the physics laboratory? What is the larger S&T framework within which research (jugaad or otherwise) happens? Is there a particular set of circumstances that need or demand a jugaad kind of enterprise? What happens/will happen when that framework changes, as it is changing now in the case of financial resources?

A deeper understanding of jugaad

There are no detailed accounts or histories of even the well-established examples of jugaad. How was the jugaad automobile actually made? What did it cost in a particular case and what have been the financial returns, if any? What parts were used and where were they accessed from? How much time did it take to build? Were there any run-ins with the law?

There are also many questions around the characterisation of technological jugaad that is the key conceptual frame offered in this book. If this technological jugaad inhabits a space that is coloured grey from the legal point of view, what role does the state, or certain arms of the state play? Does the state allow jugaad by turning a blind eye or because it is short on certain resources itself? Or is it possible that the state can, and in some cases, even choose to facilitate jugaad? Is it possible to develop a suggestive toolbox that will help to engage more fruitfully with the empirics and the concepts involving jugaad? How, for instance, is jugaad different from an invention; how is it different from a hobby, not to mention bricolage and 'user-driven' innovation even though we have seen many overlaps and similarities?

There is also the question, for instance, of the cultural and economic context of jugaad. Is it possible that a disproportionately larger play is being assigned to the economic context here? While jua kali and jugaad are clearly driven by a lack of resources that is characteristic of the poverty in Kenya and India respectively, might something else be playing out in a rich, first world context of the French bricolage? Can jugaad be thought of beyond 'lack of resources'; can there be a non-economic theory of jugaad? Could there be other cultural/social reasons for innovation, indeed for jugaad? These would all be very interesting questions to engage with indeed.

Etymological explorations

An exploration into the etymology of jugaad and the family of words it relates to would give more insights and uncover other relationships and interpretations. One such word in Hindi is *jugat* – translated in one particular dictionary (Bahri, 2010) I checked as a 'feminine noun' that means art and dexterity and where its usage in particular contexts indicates the process of manipulation. It would be an interesting exercise, at the same time, to also look for jugaad's equivalents in other parts of India – the northeast and the south for instance, where the family of languages have a very different origin. My limited explorations and asking around for a corresponding idea in the southern languages (Telugu, Tamil, Malayalam and Kannada) did not elicit the kind of spontaneous responses that jugaad manages in the northern parts. Everyone admits it exists or is bound to exist, but perhaps it is less prevalent or, then, more subtle in its manifestation.

While jugaad, we have seen, does not have an exact equivalent in English, it is noteworthy symmetry that innovation does not appear to have an equivalent in the Indian languages mentioned in the very beginning. The words and concepts that do exist are more like literal translations of innovation suggesting 'newness' or 'doing new things.' The case with invention, that close cousin of innovation, is rather different. It translates perfectly as *avishkar*, leaving us with the tantalising possibility that the culture and language of *avishkar* and jugaad never had any use of or need for what is innovation in English.

Annexures

Annexure 1

Diary notes from my first meeting with Prof CV Dharmadhikari in December 2010

The most striking thing about Dharmadhikari is that he claims (and I don't doubt it) that he along with his students fabricated a STM in their lab in the university in 1988, a couple of years after IBM and the Nobel Prize. It is an amazing achievement if it is completely true. He was invited to visit IBM where he saw that first STM and he came back and decided he'd try and make it too because it was perhaps relevant for his work.

It was much cheaper and many people did not believe him in the beginning (he kept saying that), but then students did their work, got their PhDs, have got jobs and he is happy with that – getting science out of the instrument: "It is not a question of less or more sophisticated – It is whether it works or not and in our case it works."

He took me to the lab to show me the STMs (one is built inside a fridge casing and because of the wool in the sides works well for acoustic isolation. Also that a student from the Middle East was going back home and he didn't know what to do with the fridge. That is how the fridge came into the lab, was hollowed out and the second STM was built.

He spoke of having and needing multidisciplinary skills to be able to create the instrument – information about physics, chemistry, acoustics, engineering – also that he knew many people and they came and helped him in the fabrication – those, for example, with acoustic skills and those who make springs etc. He kept emphasising that this was an extremely low cost instrument and he certainly seemed proud of it.

* * *

When he was showing the lab, I asked him whether costs were an impediment and he said no because they were all so low cost – whereas a new STM today costs something like Rs 1 crore. Now there is so much money that people are just buying STMs sometimes without even knowing how to use it and have to get help and support in training. He said there must be about 120 STMs in the country today. He also narrated the story of a lab that had bought an AFM and thought it was an STM.

***

He seemed quite interested in the work I was doing – this STS kind of thing. When we had begun in his office I gave him a considerably detailed account of STS and pushed the boundaries even further by talking of democratisation of science etc. At one moment I was wondering if I was pushing too much, but in the end I don't think that was the case. He said later that he would not have given me much time, but when he spoke to me on the phone earlier in the day he saw my excitement and I sense that he is also interested in the subject. He was clearly quite taken in with the STS kind of work – and asked just before we parted who else was doing this kind of work in India and that it was indeed interesting work.

***

I asked him if he had documented all the work – particularly of this instrument fabrication that he was doing in the lab and he said he hadn't done much. They were so busy publishing papers and doing science that he didn't have time for all this. But he has written a couple of papers and seemed to indicate that he now wants to do some documentation of what they have done. I am thinking this is a step where I can perhaps step in – offer to work for him on the documentation and maybe do some photography of his equipment – the lab and the people working in the lab.

***

Both the labs are really small – STM must've been 15'x15' and the AFM 20'X15.' All kinds of boxes and instruments lie randomly. Looks extremely messy and disorganised – also many look locally fabricated – some in plastic tiffin boxes. The walls are pasted with posters produced by students of the lab. He also said that he gets students to work on things where the earlier one has left behind, so there is some kind of continuity.

***

Actually, the explanation was interesting because he gave me some details of how the STM vibration isolation was created – that it is based on just three balls and that springs were specially fabricated and installed to isolate the equipment from vibrations. In the initial days when someone opened the door, the vibration would cause the STM to oscillate and you would only get a sine wave. Then he explained how they had created a box lined with cotton wool to ensure acoustic isolation, that someone in Pune who is the leading maker of auditoriums came with his tools to help him ensure this acoustic insulation/isolation etc.

Annexure 2

Receipts for the purchase of soldering material, shaving blades and tungsten wire used in the labs

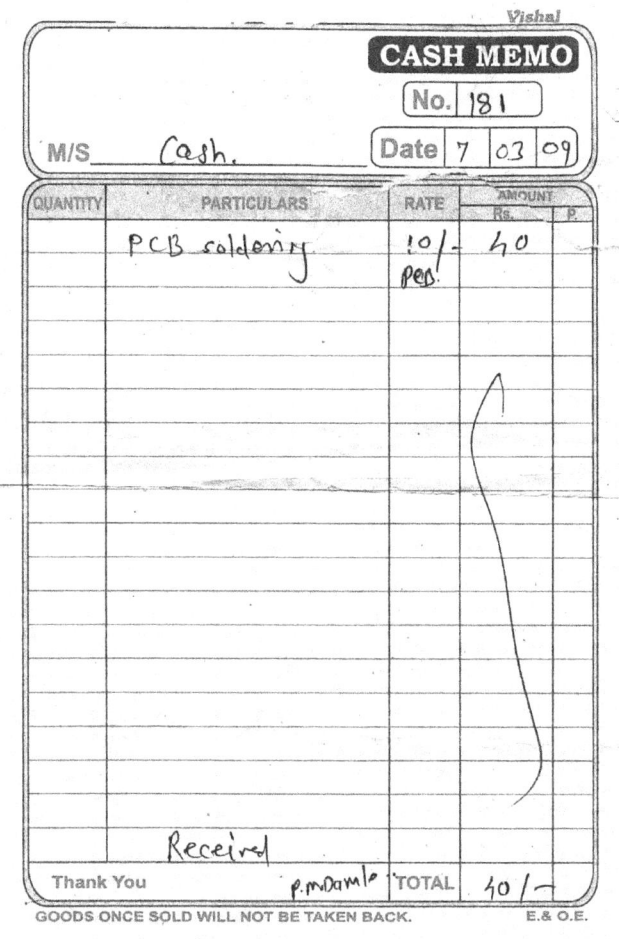

YOGESH MEDICALS

97, Ramkrishna Smriti, Pashan, Pune-21. Tel. 380614

No. **1457**

Date 8/6/03

Shri

Particulars	Rs.	Ps.
1 P S Eullette 70 clech Pulm blubs	95	00
	95	00

D. L. No. 20-10090 21-10387 20C-2007

E & O E.

Sign.

TAX INVOICE CUM GATE PASS

ORIGINAL FOR BUYER

Authorised B.
Authorised Sign...

E.L.F. FILAMENT INDUSTRIES PVT. LTD.
18, 12 Electronic Co-Op. Estate Ltd.,
Pune - Satara Road, Pune 411 009 (INDIA)
TEL. NO. : (020) 24227182, 24220478, 24221094
FAX - (020) 24227794
E.C.C. No. AAACE 4035 E XM 001 dt. 06-12-99
TIN NO. :- 27140008801/ V CST NO. :- 27140008601 / C / Dt. :1/4/2006

SR. N⁰ 000206
DATE 24/10/2008
TIME OF REMOVAL 13:45
RANGE : P IV   DIV - IX
ADDRESS : 'C' WING, ICE HOUSE, 41- A, SASOON ROAD, P...

CONSIGNEE:	CHALLAN NO. 1842	DATE 1/2008
University of Pune, Pune 411 007	ORDER NO. 81019600 DATE	
	TARIFF HEADING NO. :	
	NO. OF PACKAGES :	
CATEGORY : INDUSTRIAL CONSUMER	MODE OF TRANSPORT : POST/ COURIER / H. D./ ROAD / AIR	
NATURE OF REMOVAL : INDEPENDENT BUYER	NO. DATE :	
BILL NEGOTIATED THROUGH : Direct	NAME OF TRANSPORTER :	

DESCRIPTION OF GOODS	QUANTITY		ASSESSABLE VALUE PER UNIT		TOTAL ASSESSABLE VALUE	
	UNIT	NOS.	RS.	PS.	RS	PS.
~~Tungsten Wire~~						
Tungsten Wire	Mtrs.					
230 Micron		2	150.00		300.00	

Amount of Basic Excise Duty Payable in Words Rs.: Forty Two Only	Assessable Value 300.00
	Forwarding Charges
Amount of Cess Payable in Words Rs.: One Only	Payable Excise Duty 14% 42.00
	Payable Edu. Cess 2%
Amount of S. H. E. Cess Payable in Words Rs.: One Only	Payable H. Edu. Cess 1%
	Sub Total
Certified that the particulars given above are true & correct and the amount indicated represents the price actually charged & that there is no flow additional consideration directly/ indirectly from the buyer.	VAT 4%
TOTAL RS. IN WORDS **Three Hundred Fifty Eight]Only**	
	GRAND TOTAL 358.00

We hereby certify that our registration Certificate under Maharashtra Value Added Act 2002 is in force on the date on which sale of goods specified in which sale of goods specified in this tax invoice and that transaction of sales covered by this invoice has been effected by us in the regular course of business.
E.& O.E.

FOR E.L.F. FILAMENT INDUSTRIES PVT. LTD

Cash Received S.Y.Umaye

Annexure 3

Facsimile of Bendre and Dharmadhikari's 1988 paper, one of the first on an STM related subject to be published by the lab

JOURNAL OF OPTICS
Vol-17. (July-Sept) 1988

DESIGN, CONSTRUCTION AND CALIBRATION OF A PZT MICROMANIPULATOR FOR SCANNING TUNNELING MICROSCOPE (STM)

S. BENDRE AND C. DHARMADHIKARI

Department of Physics, University of Poona, Pune-411 007.

(Received November 17, 1987)

ABSTRACT--The operation of Scanning Tunneling Microscope (STM) depends on the ability to control and measure distances ranging from a few angstroms to several thousand angstroms. In this paper we describe the design and construction of a PZT micromanipulator for STM. The micromanipulator has been calibrated using simple interferometry and capacitance measurement principles.

1. INTRODUCTION

The Scanning Tunneling Microscope (STM) developed by Binning et al, is a novel tool which provides valuable information of the surface structure in real space with atomic resolution [1]. In essence, the instrument consists of a sharp metal tip, positioned at about ten angstroms above the sample surface with the help of a PZT tripod. The application of a bias voltage between the sample and the tip under investigation causes tunneling current to flow from negative to positive electrode. The tip is scanned over the surface by applying voltages to X and Y piezos while keeping the tunnel current constant. Since the tunnel current has an exponential dependence on the tip to the sample distance, keeping constant current is equivalent to keeping fixed distance between the tip and the sample. This is carried out by comparing tunnel current at any instant with the set value and applying voltage to Z-piezo which is proportional to the difference between instantaneous and set value through a feedback loop involving a PID controller [2]. The voltage to the Z-piezo then is a measure of the surface topography. The resolution in this case, to a first approximation, would depend on the tip radius and the distance between the tip and the sample.

One of the important aspects of STM construction is the control of the position of the tip with respect to the sample. It requires the control of the distances of the order ranging from several angstroms to a few angstroms. General approach so far adapted involves two step manipulators consisting of a coarse and fine positioning. The coarse positioning involves moving the probe towards the sample from a distance of several millimeters to a few microns. The fine positioning involves moving the sample from a few microns to the tunneling distances, which are typically of the order of few angstroms. The purpose of this paper is to report the design and construction of a micromanipulator that can be used for controlling the distances ranging from a few angstroms to few thousand angstroms. As a preliminary attempt this micromanipulator has been calibrated using simple interferometry and capacitance measurement principles.

2. EXPERIMENTAL ARRANGEMENT

A PZT micromanipulator was fabricated by connecting three piezoelectric bars of size 18 X 3 X 3 mm orthogonal to each other with Araldite. The bars were cut from PZT-4 disc (diameter 50mm X thickness 3mm) which had silvered coating on either side and poling perpendicular

Annexure 4

Rajendra Kshirsagar's presentation during the seminar in March 2011 to felicitate Dharmadhikari

Friends, I am grateful for giving me this opportunity for expressing my thoughts. The great British Somerset Maugham once said that there are three rules for writing a novel – unfortunately nobody knows what they are. So, I think when a fresh MSc student joins for a PhD, for getting a PhD – it's such an enormous task that he doesn't know what the rules are and for that reason the role of the guide is one of the most important things in the life span of the PhD. I consider myself very fortunate that I had Prof Dharmadhikari as a guide for my PhD. When I joined the lab in 1995–96, [the] STM still was somewhat of a novelty in the country. (. . .) People (. . .) did not believe that the STM really worked and (. . .) whenever we used to go for conferences, they used to ask (. . .) what is happening.

So I consider myself to be very fortunate for getting the opportunity to work on a home made STM because when you work on a home made instrument, you have to learn all the parts and learn the techniques inside out. This has helped immensely in my research, [including] in my post doc research. So at that point of time, getting the opportunity to work on such a wonderful instrument – I consider it a great opportunity. Apart from that, I learnt a lot from Prof Dharmadhikari during my association with him. He has a keen sense of experimental insight, which is quite rare and he has also the knack of finding unusual problems – problems which are not routine. So he has a knack of attacking interesting and scientifically valuable problems.

Apart from the academic qualities, I also learnt a lot of qualities outside academics. For e.g. as Prof Jog and Prof Kshirsagar mentioned, he has a very keen sense of humour. This has helped us in many situations. I found that my own sense of humour got a boost (laughter in the audience) after I joined the lab – much to the dismay of my lab colleagues sometimes (laughter continues). This sense of humour, he used to (. . .) enliven the atmosphere, but during our PhD he also used it in some other ways. For example, when I was not working, which is more often than I would like to admit (laughter), he would not tell me straight away to do something or to work or whatever it was. He would tell me a very funny story, which had a moral and that moral exactly applied to my situation. I got the message very clearly (laughter). So I found this quality is very rare and is really helped me during my PhD.

I take this opportunity [to express] my heart felt thanks for all the things he has done for me. And it is because of him that my stay during the PhD in the lab has been very satisfying. And on occasions like this there is a touch of sadness but I still wish him a fruitful life and my best wishes will always remain with him.

Thank you very much.

Annexure 5

Extract of interview with Shirshendu
Dey, who completed his PhD under
Dharmadhikari's supervision in 2011*

. . .

(Q) From a student and learning point of view, do you feel that (. . .) because
  you made your own instrument and did your own science in that sense,
  are you better equipped in some way? (. . .) Is there a qualitative difference in your learning, or in the way that you now look at instruments
  and understand science because you came through a certain channel or a
  methodology of doing science?
(Dey) Definitely.

(Q) And what would that kind of experience be?
(Dey) Okay. I will go by my experiences. I have designed this set up (. . .)
  [and] I know almost anything about it. Just by seeing the signals and
  the noise I can trace which wire is creating the problem – that I can
  make out.

(Q) That is because you have put it together?
(Dey) Ya. I know what has gone into it. So, I am now working with a
  US$70,000 worth machine, probably US$75,000 worth instrument (. . .).

(Q) Hmm. Hmm.
(Dey) [A] STM. There we do the same – sometimes same types of (. . .) problems come. Then within like a day I can [figure it out]: we have seen this
  thing, this might be from this, let's tap this wire, this grounding might be
  faulty. Change that wire and the problem [is sorted out]. But in the same
  machine, [another] guy was working for almost one and half year and he has
  been using the machine, same machine, for some other studies. He was also
  seeing the same thing but he didn't have this electronics background of the
  development. He couldn't figure it out.

(Q) So, then would you also say that you actually require – it is a different
  kind of training and a different kind of experience and experimentation

and you require. Do you feel that somebody who is coming through the development channel is a better person to work on that equipment?

(Dey) Okay. [It's] (. . .) a very tricky question to answer in a one sentence. (. . .) Let's (. . .) look at the group into which I have been appointed [at the University of California]. So (. . .) why have they appointed me? And why have they appointed this other guy? Like, (. . .) [my] boss – [he has] (. . .) seen everything in the world. So (. . .) even if you are buying a 75000 [worth equipment], it will show some problems – these things (. . .) are complicated. So, he has a person who has some different background [and] (. . .) there comes my job.

(Q) You have a different set of expertise?

(Dey) Ya, I have a different set of expertise. So, why [has] he put me in that group? To solve this kind of thing.

(Q) It's like creating the team with complementary experience and expertise.

(Dey) Correct. Ya. Ya. Ya. Because the group like I am in – it's in the Chemistry department. And how much [do] I know [about] Chemistry? Nothing. Next to nothing. That's why I am being appointed. (. . .) So, you need people of different expertise to move your projects.

(Q) [There is an opinion] that developing equipment is like wasting time, like reinventing the wheel. When you can, when you have the money and you can get equipment, why don't you get it and do experimentation? Why do want to put people through this process of making your equipment? Why do you want to re-invent the wheel? (. . .) Do you feel it is reinvention of the wheel or do you feel that it is, like you are saying, you create certain expertise which is complimentary or can be made complimentary in a research group? Would you rather have equipment that you can do the experiment or do you feel this training of a particular kind is as valid and is needed. (. . .) Supposing the lab gets Rs. 30–40 crores (. . .) and they just buy the equipment and do science? Then this kind of science, this kind of instrumentation, does it become irrelevant? And do you feel it should become irrelevant or do you feel there is a value to this kind of training and instrumentation building?

(. . .)

(Dey) My personal feeling is that you do need (. . .) to know how to develop instruments. Nowadays, the instruments that you are getting, if you don't know and don't have a basic training on the usage, you can never get a good result out of it. That's for sure. (. . .) There are many groups in India [with] (. . .) several STMs lying there. These

are very good instruments, [but why isn't there] (. . .) great productivity? (. . .) You should be getting, like, a flourish of results from
those labs. But I haven't seen a very good paper [or a] very good
STM image from those. Why? It is not that people are not trained or
[are] not well versed with scientific knowledge. It's the instruments.
These instruments do have problems. (. . .) We face this problem. The
instrument was a fabulous instrument, but you need to understand. If
it is showing something which it should not be showing. But why it is
showing? There is something wrong with it. Right?

(Q) Can't you get the engineer from the company and solve it?

(Dey) No.

(Q) Why?

(Dey) He will come. He will say, okay – he will sit with you one day, two
days. But you are running the experiment day and night. (. . .) Also call-
ing an engineer (. . .) costs money, time. You need to understand these
things. Like in US, where if you call the guy he will come within one
day. [Even] then (. . .) at the end of the day you need to understand your
instrument.

(Q) And of course you are paying him and there are other costs involved.

(Dey) Correct. There are huge costs. And that's possible in US. In California
that is possible, but in India? (. . .) How can you bring that engineer? You
have to depend on a person. Maybe their service station is in Bangalore.
(. . .) Everywhere costs are involved . . . and time! More important is
time. (. . .) You will see that the best groups in the world, they all have
their own set ups. You take anybody. (. . .) Don't go by my comments –
you take the literature, do a literature review. [Take the] last five years.
80% to 90% you will see that people have done in home-made set ups.
Are they fools? Or they are not intelligent enough? Or do they have a
lack of funds? Like, why? (. . .) For scanning probe microscopy, these are
open end instruments because you can do a hell of a lot of things. (. . .)
And no company can give you (. . .) what you want. Like, at the end of
the day why you are there? You want to innovate. (. . .) And it is good to
have your own set up.

Note

* He had recently got a post-doctoral appointment at the University of California,
USA

Annexure 6
A brief note on nanoscience and technology in India

The modern legend of nanotechnology has been built, quite interestingly, by mobilising the deep past. The centuries-old Lycurgus Cup is one of the prominent objects showcased to illustrate both the beauty and the possibilities that nanotechnology offers. Made of a special kind of glass, dichroic glass, it changes colour when it is held up to light. The opaque green when seen in regular light turns as if "by magic" (Rao & Kaur, 2011) to a glowing translucent red when light is made to shine through it. This is on account of small amounts of nanoscale (what was earlier called colloidal) gold and silver present in it. Another famous object is the Damascus Sword that was first made in the 8th century and was forged from Wootz steel made in India, reportedly, as early as 300 BC. The secret behind the sword's famed strength and sharpness were decoded recently by scientists at the Technical University in Dresden, Germany. Electron microscope studies of the blade of one such sword made in the 17th century revealed the presence of tiny nanowires and the first ever carbon nanotubes present in steel (Rao & Kaur, 2011). In another more relevant Indian context, there has been a prominent narrative and much recent scientific work to claim overlaps between the Ayurveda and Siddha systems of medicine and nanotechnology. The *bhasmas* (metallic oxides or ashes) that are used to treat a range of conditions in the *Rasashastra* tradition are considered effective precisely for the reason that they contain metals at the nanoscale (Umrani, Agrawal, & Paknikar, 2013; Valiathan, 2006).

Not a single week passes without reports of nanoscience and technology-related scientific breakthroughs, technological outputs and possible solutions from India's current-day scientific establishment as well. Sample this from half-a-year's reporting in *The Hindu*, one of India's leading newspapers that provides prominent space to new R&D in S&T: development by the Institute of Advanced Study in Science and Technology, Guwahati, of a smart bandage material using nanosized graphene oxide that can heal wounds better and faster (R. Prasad, 2018c); development of a carbon nano-tube based electronic skin (e-skin) that "may find applications in flexible electronics and medical diagnostics" (R. Prasad, 2017) and of a nano-silver impregnated novel composite that "keeps tomatoes fresh for 30 days" (R. Prasad, 2018b), both at the Indian Institute of Technology-Hyderabad and the use of a magnetic nanofluid emulsion by scientists at the Indira Gandhi Centre for Atomic Research, Kalpakkam, near Chennai, to develop an "inexpensive, highly sensitive optical probe that can almost instantaneously detect the presence of urea" (R. Prasad, 2018a).

All of this has undoubtedly been possible because of substantial support and incentives from within the policy framework, which has itself been driven to a certain extent by the fear of missing the 'nano-bus' as was articulated in 2011 by CNR Rao, then scientific advisor to India's Prime Minister ("India in danger of missing 'nano bus': PM's Scientific advisor," 2011). The Government of India (GoI) launched a Nano Science and Technology Initiative (NSTI) in the year 2001–2002 through its Department of Science and Technology (DST). An estimated Rs. 125 crore (approx. US$25 million)[1] were spent as part of the initiative for the period 2002–2007. This was followed by a much larger program called NanoMission[2] where the government committed (again through the DST) an unprecedented Rs. 1000 crores (approx. US$250 million)[3] for a five-year period starting May 2007 (DST, 2007). An extended Phase 2 of NanoMission was subsequently approved in 2012 with a budgetary allocation of Rs. 650 crores (approx. US$135 million)[4] (PIB, 2014).

The DST has over the years supported a number of programs and projects in education related to NS&T, to scientific research in laboratories and institutions, to collaborations between academia and industry and to help organise a series of national- and international-level conferences that have brought together the best minds in the field. A number of other central government institutions – the Department of Electronics and Information Technology, Department of Biotechnology, Council of Scientific and Industrial Research, Defence Research and Development Organisation, Indian Council for Medical Research, Indian Space Research Organisation and the Department of Atomic Energy – have also independently invested substantial resources in NS&T research in the last decade (Bhattacharya, Pushkaran, Shilpa, & Bhati, 2012).

The number of NS&T related publications by Indian scientists, an important marker of output and efficiency, has also increased steadily – from 355 in 2001, to 1413 in 2006, to 4911 in 2012. In 2011, India was ranked sixth in terms of its global contribution to NS&T publications (6% of global output), a significant improvement from its 13th rank in 2000 when its contribution was 2% of the global output. International patent applications in nanotechnology by Indian assignees too have increased correspondingly – from 7 in the year 2001 to 32 in 2006, to 114 in 2010.[5]

The popularity of the field can be also be ascertained from the increasing number of conferences, workshops and seminars that are being held in India. Conferences like the International Conference on Nano Science and Technology (ICONSAT) that are held every other year (2016 in Pune; 2014 in Mohali; 2012 in Hyderabad; 2010 in Mumbai) and Bangalore India Nano (BIN),[6] which is held every year, have become prominent features of the conference calendar in the country and continue to be extremely popular as attested by the delegates attending – over 800 at ICONSAT 2012 and more than 600 at BIN in 2016.

There are also at least two (non-academic) periodicals that are devoted exclusively to issues of NS&T: *Nano Digest*,[7] which was started in 2009 and is privately published, and *Nanotech Insights*[8] that was started in 2010 by the Centre for Knowledge Management of Nanoscience and Technology (CKMNT), which was set up with partial financial support from NanoMission. An increasing interest is to be seen in industry as well with one preliminary study published in 2012 suggesting that 500 companies (100 being in the area of pharma/nano bio-pharma) were working on nanotechnology and related products, and that about 50 companies had already commercalised such products (Purushotham, 2012).

One very important detail in the policy context here is related to the funding that has been made available. Of the Rs. 1000 crores (approx. US$250 million) allocated for NS&T research and development when NanoMission began in 2007, nearly 50% (Rs. 500 crores; approx. US$125 million) had not been spent when the first phase of the program came to a close in May 2012 (Bhattacharya et al., 2012; Kumar, 2014). The challenge in India, it emerges then, is not that of limited financial resources but of multiple deficits in the capacity of the research establishment that includes colleges, universities, government institutions and industries to utilise the resources that are already available.

Notes

1 1 US$ = Rs. 50 (approx. exchange rate in 2002); as per the current exchange rate (June 2018), 1 US$ = Rs. 70 (approx.).
2 For more details see the NanoMission website. Retrieved from www.nanomission.gov.in.
3 1 US$ = Rs. 40 (approx. exchange rate in 2007).
4 1 US$ = Rs. 50 (approx. exchange rate in 2012).
5 For a more detailed bibliometric analysis related to Indian NS&T publications, citations and patenting see Bhattacharya, Shilpa, Pushkaran (2012), Patel (2012), Patel and Chander (2012) and Purushotham (2013).
6 Retrieved from www.bangaloreindianano.in/.
7 Retrieved January 10, 2014, from www.nanodigestmag.com/.
8 Retrieved January 10, 2014, from www.ckmnt.com/newsletter.html.

Annexure 7
List of interviews

	Name	Institution	Date (place)
1	Dey, Subhendu	*Formerly*, Department of Physics, University of Pune, Pune	20 July 2011 (Pune)
2	Dharmadhikari, C.V.	Department of Physics, University of Pune, Pune; *Currently*, Indian Institute of Science Education and Research (IISER), Pune	13 December 2010, 02 March 2011, 24 April 2011 (Pune)
3	Dharmadhikari, Mona		25 April 2012 (Pune)
4	Iyyer, S. B.	*Formerly*, Department of Physics, University of Pune, Pune; *Currently*, Ahmednagar College, Ahmednagar	24 October 2011 (Pune)
5	Krishnan, Rishikesha	Indian Institute of Management Bangalore, Bengaluru	06 July 2011 (Bengaluru)
6	Kshirsagar, Rajendra	*Formerly* Department of Physics, University of Pune, Pune	21 October 2011 (Pune)
7	Mashelkar, R. A.	National Chemical Laboratory, Pune	08 June 2011 (Pune)
8	Patil, Shivaprasad	*Formerly* Department of Physics, University of Pune, Pune; *Currently* Indian Institute of Science Education and Research (IISER), Pune	09 March 2011 (Pune)
9	Patil, Sumati	Department of Physics, University of Pune, Pune	24 October 2011 (Pune)
10	Pillai, Vijaymohanan	Central Electrochemical Research Institute, Karaikudi	04 April 2012 (Goa)
11	Rao, C. N. R	Jawaharlal Nehru Centre for Advanced Scientific Research (JNCASR), Bengaluru	03 August 2012 (Bengaluru)

	Name	Institution	Date (place)
12	Raychaudhuri, Arup	SN Bose National Centre for Basic Sciences, Kolkata	22 January 2012 (Hyderabad)
13	Sastry, Murali	*Formerly* Tata Chemicals Ltd. Pune; *Currently*, CEO, IITB-Monash Research Academy, Mumbai	14 March 2012 (Gurgaon), 24 July 2012 (Pune)
14	M1	University of Pune, Pune	05 April 2011, 08 April 2011 (Pune)

References

Abraham, I. (2000). Landscape and postcolonial science. *Contributions to Indian Sociology, 34*, 163–187.

Abraham, I. (2006, January 21). The contradictory spaces of postcolonal techno-science. *Economic and Political Weekly*, 210–217.

Abrol, D. (2013). Where is India's innovation policy headed? *Social Scientist, 41*(3–4), 65–80.

Abrol, I. P., & Sangar, S. (2006). Sustaining Indian agriculture – conservation agri-culture the way forward. *Current Science, 91*(8), 1020–1025.

Aiyar, M. S. (2015). Celebrating diversity. *Seminar, 667*, 52–56.

Anand, U. (2012). *Apex court wants jugaad out, asks states for reports*. Retrieved from www.indianexpress.com/news/apex-court-wants-jugaad-out-asks-states-for-report/988890/

Anderson, R. S. (2011). *Nucleus and nation: Scientists, international networks, and power in India* (First Indi). Chicago, IL and New Delhi: The University of Chi-cago Press, Supernova Publishers & Distributors Pvt Ltd.

Arnold, D. (2013). Nehruvian science and postcolonial India. *Isis, 104*, 360–370.

Bahri, H. (2010). *Learners' Hindi-English dictionary*. New Delhi: Rajpal & Sons.

Baird, D. (2004a). *Navigating nano through society*. University of South Carolina.

Baird, D. (2004b). *Thing knowledge: A philosophy of scientific instruments*. Berke-ley: University of California Press.

Balakrishnan, P. (Ed.). (2011). *Economic reforms & growth in India – essays from the economic and political weekly*. Hyderabad: Orient BlackSwan.

Balaram, P. (1999). Sanctions. *Current Science, 76*(9), 1171.

Balaram, P. (2012a). Innocence and sophistication: Users and equipment. *Current Science, 102*(9), 1241–1242.

Balaram, P. (2012b). Tools as drivers of science. *Current Science, 103*(12), 1383–1384.

Banerjee, P. (Ed.). (2009). *India science and technology 2008: Summary*. New Delhi: CSIR-National Institute of Science, Technology and Development Studies (NISTADS).

Bendre, S. (1987). *Design of micromanipulator for STM* (MSc thesis). Department of Physics. University of Pune, Pune.

Bendre, S., & Dharmadhikari, C. (1988, July–September). Design, construction and calibration of a PZT micromanipulator for Scanning Tunneling Microscope (STM). *Journal of Optics*, *17*, 67–70.

Bhattacharya, S., Pushkaran, J. A., Shilpa, & Bhati, M. (2012). *Knowledge creation and innovation in nanotechnology: Contemporary and emerging scenario in India*. New Delhi: CSIR-National Institute of Science, Technology and Development Studies (NISTADS).

Bhattacharya, S., Shilpa, & Pushkaran, J. A. (2012, July). *Nanotechnology research and innovation in India: Drawing insights from bibliometric and innovation indicators*. CSIR-NISTADS Policy Brief – II. New Delhi: CSIR-National Institute of Science Technology and Development Studies (NISTADS).

Bijker, W. E., Hughes, T. P. & Pinch, T. (Eds.). (1987). *The Social Construction of Technological Systems. New Directions in the Sociology and History of Technology*. Cambridge, Mass.: The MIT Press.

Bijker, W. E. (1995a). *Democratisering van de Technologische Cultuur (Inaugurele Rede)*. Maastricht: Universiteit Maastricht.

Bijker, W. E. (1995b). *Of bicycles, bakelites, and bulbs: Toward a theory of sociotechnical change: Inside technology*. Cambridge, MA: The MIT Press.

Bijker, W. E. (2006). The vulnerability of technological culture. In H. Nowotny (Ed.), *Cultures of technology and the quest for innovation* (pp. 52–69). New York: Berghahn Books.

Binnig, G., & Roehrer, H. (1986). *Scanning tunneling microscopy – from birth to adolescence*. Nobel Lecture.

Birtchnell, T. (2011). Jugaad as systemic risk and disruptive innovation in India. *Contemporary South Asia*, *19*(4), 357–372.

Bound, K., & Thornton, I. (2012). *Our frugal future: Lessons from India's innovation system*. London: Nesta.

Cappelli, P., Singh, H., Singh, J., & Useem, M. (2011). *The India way – how India's top business leaders are revolutionizing management*. Boston, MA: Harvard Business Review Press.

Chacko, P., Noronha, C., & Agrawal, S. (2010). *Small wonder – the wonder of the nano*. Chennai: Westland Ltd.

Chaki, N. K., Singh, P., Dharmadhikari, C. V., & Vijayamohanan, K. P. (2004). Single-electron charging features of larger, dodecanethiol-protected gold nanoclusters: Electrochemical and scanning tunneling microscopy studies. *Langmuir*, *20*(23), 10208–10217.

Chakrabarti, P. (2004). *Western science in modern India: Metropolitan methods, colonial practices*. Ranikhet: Permanent Black.

Chaudhary, M. V. (2002). *Development of dynamic Michelson interferometry for the callibration of intertial translator* (M Phil thesis). Department of Physics. University of Pune, Pune.

Chaudhary, M. V. (2011). *Scanning probe microscopic investigations of metallic nanostructures: Electron transport, charge storage and related processes* (Ph D thesis). Department of Physics. University of Pune, Pune.

Chopra, K. L. (2016). From physics to engineering physics. In P. Ghosh & B. Raj (Eds.), *The mind of an engineer* (pp. 297–308). Springer.

Collins, H. M. (1985). *Changing order: Replication and induction in scientific practice*. London: Sage.

Collins, H. M. (1987). Expert systems and the science of knowledge. In W. E. Bijker, T. P. Hughes, & T. Pinch (Eds.), *The social construction of technological systems: New directions in the sociology and history of technology* (pp. 329–348). Cambridge, MA: The MIT Press.

Colton, R. (2005). *Letter of recommendation*.

Constant, E. W. (1987). The social locus of technological practice: Community, system, or organisation? In W. E. Bijker, T. P. Hughes, & T. Pinch (Eds.), *The social construction of technological systems: New directions in the sociology and history of technology* (pp. 223–242). Cambridge, MA: The MIT Press.

Dambe, A. T. (1995). *Development and application of electronic system for scanning tunneling microscopy* (M Phil thesis). Department of Physics. University of Pune, Pune.

Datar, S. (2004). *Some aspects of investigation of nanostructures using Scanning Tunneling Microscopy (STM)/Atomic Force Microscopy (AFM) and related techniques* (Ph D thesis). Department of Physics. University of Pune, Pune.

Datar, S., Patil, S., Iyyer, S., & Dharmadhikari, C. (2004). Scanning force microscopy and amplitude versus distance measurements on single-crystal oxide surfaces. *Surface and Interface Analysis, 36*, 213–219.

Datta, P. (2010). The dividends of innovation (Editorial) In P. Datta (Ed.) *A Case Study Special on Innovation – Making Aspirations Count*, (p. 4). New Delhi: BusinessWorld.

Deloitte. (2012). *Indian higher education section – opportunities aplenty, growth unlimited*. Retrieved from www2.deloitte.com/content/dam/Deloitte/in/Documents/IMO/in-imo-indian-higher_education_sector-noexp.pdf

Dey, S. (2010). *Design and development of photo emitting STM to study the optical properties of individual nanostructures* (Ph D thesis). Department of Physics. University of Pune, Pune.

Dey, S., Pethkar, S., Adyanthaya, S. D., Sastry, M., & Dharmadhikari, C. V. (2008). New approach towards imaging λ-DNA using Scanning Tunneling Microscopy/Spectroscopy (STM/STS). *Bulletin of Material Science, 31*(3), 309–312.

Dharmadhikari, C. V., & Gomer, R. (1984). Diffusion of hydrogen and deuterium on the (111) plane of tungsten. *Surface Science, 143*, 223–242.

Douglas, M. (1966). *Purity and danger: An analysis of concept of pollution and taboo*. Routledge and Keegan Paul: Abingdon.

DST. (2007). *Report of the working group on DST – eleventh five year plan 2007–12*. New Delhi: DST.

Eigler, D. M., & Schweizer, E. K. (1990). Positioning single atoms with a scanning tunneling microscope. *Nature, 344*, 534–526.

Fagerberg, J., Fosaas, M., & Sapprasert, K. (2012). Innovation: Exploring the knowledge base. *Research Policy, 41*, 1132–1153.

Felt, U., Fouche, R., Miller, C. A., & Smith-Doerr, L. (2017). Introduction to the fourth edition of the handbook of science and technology studies. In U. Felt, R.

Fouche, C. A. Miller, & L. Smith-Doerr (Eds.), *The handbook of science and technology studies* (4th ed., pp. 1–26). Cambridge, MA: The MIT Press.

Fischer, M. M. J. (2007). Culture and cultural analysis as experimental systems. *Cultural Anthropology, 22*(1), 1–65.

Fortun, K. (2001). *Advocacy after Bhopal: Environmentalism, disaster, new global orders*. Chicago, IL: The University of Chicago Press.

Freeman, C. (1987). *Technology policy and economic performance: Lessons from Japan*. London: Pinter Publishers.

Gadagkar, R. (2015). Solve local problems. *Nature, 521*, 153.

Ganesh, K. N. (2015). Connect research with education. *Nature, 521*, 154.

Garcia, R., & Calantone, R. (2002). A critical look at technological innovation typology and innovativeness terminology: A literature review. *The Journal of Product Innovation Management, 19*, 110–132.

Geertz, C. (1973). *The interpretation of cultures*. New York: Basic Books.

Gibbons, M., Limoges, C., Nowotny, H., Schwartzman, S., Scott, P., & Trow, M. (1994). *The new production of knowledge: The dynamics of science and research in contemporary societies*. London: Sage.

Giridharadas, A. (2010). A winning formula for hard economic times. *The New York Times*. New York.

Godbole, V. P., Sumant, A. V., Kshirsagar, R. B., & Dharmadhikari, C. V. (1997). Evidence for layered growth of (100) textured diamond films. *Applied Physics Letters, 71*(18), 2626–2628.

GoI. (1958). *Scientific policy resolution*. New Delhi: Govt. of India.

GoI. (2003). *Science and technology policy*. New Delhi: Govt. of India.

Govindarajan, V., & Trimble, C. (2012). *Reverse innovation: Create far from home, win everywhere*. Cambridge, MA: Harvard Business Review Press.

Gregson, N., & Crang, M. (2010). Materiality and waste: Inorganic vitality in a networked world. *Environment and Planning, 42*(5), 1026–1032.

GTZ. (2000). Ecosan-closing the loop in wastewater management and sanitation. In C. Werner, J. Schlick, G. Witte, & A. Hilderbrandt (Eds.), *Ecosan-closing the loop in wastewater management and sanitation* (p. 327). Bonn, Germany: GTZ GmbH.

Gupta, A. (2013a). Tapping the entrepreneurial potential of grassroots innovation. *Supplement to the Stanford Social Innovation Review*, 18–20.

Gupta, A. (2013b, July 1). The grassroots innovators. *Mint*.

Habib, S. I., & Raina, D. (2007). Copernicus, Columbus, colonialism, and the role of science in nineteenth-century India. In S. I. Habib & D. Raina (Eds.), *Social history of science in colonial India* (pp. 229–252). Oxford: Oxford Univeristy Press.

harma, A. (2008). Portrait of a science teacher as a bricoleur: A case study from India. *Cultural Studies of Science Education, 3*, 811–841. doi:10.1007/s11422-008-9120-2

Hazarika, S. (1987). *Bhopal, the lessons of a tragedy*. New Delhi: Penguin Books.

Henke, C. R., & Gieryn, T. F. (2008). Sites of scientific practise: The enduring importance of place. In E. J. Hackett, O. Amsterdamska, M. Lynch, & J. Wajcamn (Eds.), *The handbook of science and technology studies* (3rd ed., pp. 353–376). Cambridge, MA: The MIT Press.

IANS. (2016). *"Technology Vision 2035": India bets on technology to overcome challenges*. Retrieved June 26, 2016, from www.financialexpress.com/article/industry/tech/technology-vision-2035-india-bets-on-technology-to-overcome-challenges/192653/

Ila, H. (2015). Support the bulk of students. *Nature, 521,* 152.

India in danger of missing "nano bus": PM's scientific advisor. (2011). Retrieved from www.deccanherald.com/content/174061/india-danger-missing-nano-bus.html

Iyyer, S. B. (1994). *Development and application of electronic control system for scanning tunneling microscope* (M Phil thesis). Department of Physics. University of Pune, Pune.

Iyyer, S. B. (2006). *Design, development and application of Scanning Tunneling Microscopy (STM) techniques for nanolithography and nanofabrication* (Ph D thesis). Department of Physics. University of Pune, Pune.

Jasanoff, S. (1994). *Learning from disaster: Risk management after Bhopal.* Philadelphia: University of Pennsylvania Press.

Jolly, M. (2009). *The Jugaad country*. Retrieved from www.vccircle.com/columns/the-jugaad-country

Jugaad in not innovation: PC. (2012, August 8). *The Indian Express.* Pune.

Kaiser, D. (Ed.). (2005). *Pedagogy and the practice of science.* Cambridge, MA: The MIT Press.

Kalam, A. P. J., & Rajan, Y. S. (1998). *India 2020: A vision for the new millenium.* New Delhi: Penguin Books.

Kapila, U. (2010). Assessment of the growth experience: Poverty, unemployment and inflation. In U. Kapila (Ed.), *Indian economy since independance* (21st ed., pp. 857–906). New Delhi: Academic Foundation.

Kaplinsky, R. (2009). *Schumacher meets Schumpeter: Appropriate technology below the radar* (No. IKD Working Paper No. 54). The Open University.

Khandekar, A., Beumer, K., Mamidipudi, A., Sekhsaria, P., & Bijker, W. E. (2017). STS for development. In C. A. Miller, U. Felt, R. Fouché, & L. Smith-Doerr (Eds.), *The handbook of science and technology studies* (pp. 665–693). Cambridge, MA: The MIT Press.

Kline, R., & Pinch, T. (1996). Users as agents of technological change: The social construction of the automobile in the rural United States. *Technology and Culture, 37*(4), 763–795.

Knorr Cetina, K. D. (1995). Laboratory studies: The cultural approach to the study of science. In S. Jasanoff, G. E. Markle, J. C. Petersen, & T. Pinch (Eds.), *Handbook of science and technology studies* (pp. 140–166). Thousand Oaks: Sage.

Kolekar, S. (2013). *Study of electron transport across nanostructures using combination of scanning tunneling/atomic force microscopy and related techniques* (Ph D thesis). University of Pune. Pune.

Krishna, V. V. (2013). *Changing social relations between science and society: Contemporary challenges*. Working Paper Series. Paris: Foundation Maison des sciences de l'home.

Krishnan, R. T. (2010). *From Jugaad to systematic innovation – the challenge for India.* Bangalore: The Utprerka Foundation.

Kumar, A. (2014). *Nanotechnology development in India: An overview* (RIS Discussion Papers). New Delhi.

Kumar, A., Pattarkine, M., Bhadbhade, M., Mandale, A. B., Ganesh, K. M., Datar, S. S., . . . Sastry, M. (2001). Linear superclusters on colloidal gold particles by electrostatic assembly on DNA templates. *Advanced Materials, 13*(5), 341–344.

Latour, B., & Woolgar, S. (1986). *Laboratory life: The social construction of scientific facts*. Princeton, NJ: Princeton University Press.

Levi-Strauss, C. (1962, translated 1966). *The Savage Mind. The Nature of Human Society Series*. Oxford: Oxford University Press. English translation 1966 by George Weidenfield and Nicholson Ltd.

Levi-Strauss, C. (1966). *The savage mind: The nature of human society series*. Oxford: Oxford University Press.

Leydesdorff, L. (2005). The triple helix model and the study of knowledge-based innovation systems. *International Journal of Contemporary Sociology, 42*(1).

Leydesdorff, L., & Etzkowitz, H. (1998). The triple helix as a model for innovation studies. *Science and Public Policy, 25*(3), 195–203.

Lin, C. H. (2009). The silenced technology: The beauty and sorrow of reassembled cars. *East Asian Science, Technology and Society: An International Journal, 3*(1), 91–131. doi:10.1007/s12280-009-9088-3

Livingstone, D. (2003). *Putting science in its place: Geographies of scientific knowledge*. Chicago, IL: The University of Chicago Press.

Lorenz-Meyers, D. (2012). Locating excellence and enacting locality. *Science Technology and Human Values, 37*(2), 241–263. doi:10.1177/0162243911409249

Lundvall, B-A. (Ed.). (1992). *National systems of innovation: Towards a theory of innovation and interactive learning*. London: Pinter Publishers.

Madhavan, N. (2013). *Nano: The blemish on Ratan Tata's otherwise brilliant run.* Retrieved from http://businesstoday.intoday.in/story/tata-nano-a-blemish-on-ratan-tata-brilliant-record/1/191897.html

Maikhuri, R. K., Negi, V. S., Rawat, I. S., Sahani, A. K., Sundriyal, R. C., & Dhyani, P. P. (2015). Traditional agriculture systems – meeting report. *Current Science, 108*(9), 1581–1583.

Mallick, S. (2014). The realm of commodified technoscience. *Seminar, 654*, 32–42.

Mani, S. (2013). The science, technology and innovation policy: An assessment. *Economic and Political Weekly, 48*(10), 16–19.

Manupriya. (2011). Science funding: Budget 2011–12. *Current Science, 100*(7), 964.

Marcelle, G. (2017). Science, technology and innovation policy that is responsive to innovation performers. In S. Kuhlmann & G. O. Matamoros (Eds.), *Research handbook on innovation governance for emerging economies – towards better models*. Edward Elgar Publishing Ltd.

Mashelkar, R. A. (2011a). India @ 75: An innovation superpower? *CSIR News, 61*(19–20), 225–232.

Mashelkar, R. A. (2011b). *Reinventing India*. Pune: Sahyadri Prakashan.

McLain, S. (2013). *Why the world's cheapest car flopped.* Retrieved from http://online.wsj.com/news/articles/SB10001424052702304520704579125312679104596

Menon, M. G. K. (1982). Basic research as an integral component of a self-reliant base of science and technology. In Anon. (Ed.), *The shaping of Indian science – Indian science congress association presidential addresses – Vol. III: 1982–2003* (Vol. III, pp. 1271–1303). Hyderabad: Universities Press.

Mishra, A. (2009). The place and space of research work: Studying control in a bioscience laboratory. In S. Bauer & A. Wahleberg (Eds.), *Contested categories: Life sciences in society*. Denmark: Ashgate.

The Missing Revolution. (2012, February 28). *Mint*. Hyderabad.

Mody, C. C. M. (2004). How probe microscopists became nanotechnologists. In D. Baird, A. Nordmann, & J. Schummer (Eds.), *Discovering the nanoscale* (pp. 119–133). Amsterdam: IOS Press.

Mody, C. C. M. (2005). Instruments in training: The growth of American probe microscopy in the 1980s. In D. Kaiser (Ed.), *Pedagogy and the practice of science: Historical and contemporary perspectives* (pp. 185–216). Cambridge, MA: The MIT Press.

Mody, C. C. M. (2006). Corporations, universities and instrumental communities: Commercializing probe microscopy, 1981–1996. *Technology & Culture, 47,* 56–80.

Mody, C. C. M. (2011). *Instrumental community*. Cambridge, MA: The MIT Press.

Mody, C. C. M., & Kaiser, D. (2008). Scientific training and the creation of scientific knowledge. In E. J. Hackett, O. Amsterdamska, M. Lynch, & J. Wajcman (Eds.), *The handbook of science and technology studies* (pp. 377–402). Cambridge, MA: The MIT Press.

Mondal, N. (2015). Build big physics facilities. *Nature, 512,* 155.

More, R. M. (1990). *A study of scanning tunneling microscope and development of a simple electronic control system for the same* (M Phil thesis). Department of Physics. University of Pune, Pune.

MST. (2013). *Science, technology and innovation policy* (M. of S. and Technology, Ed.). Ministry of Science & Technology, Govt. of India.

Mujumdar, P. P. (2015). Share data on water resources. *Nature, 521,* 151.

Munshi, P. (2009). *Making breakthrough innovation happen*. New Delhi: Harper-Collins Publishers.

Nanda, M. (2003). *Prophets facing backward – postmodern critiques of science and Hindu nationalism in India*. Rutgers University Press.

Narain, S. (2015). Manage waste frugally. *Nature, 521,* 155.

Narlikar, J. V. (2003). *The scientific edge – the Indian scientist from vedic to modern times*. New Delhi: Penguin Books.

Nazareth, S. (2017, November 3). Jugaad isn't the solution. *The Hindu*.

Nehru, J. (2003). Science in the service of the nation. In Anon. (Ed.), *The shaping of Indian science – Indian science congress association presidential addresses. Vol I: 1914–1947* (Vol. I, pp. 574–577). Hyderabad: Universities Press.

Nelson, R. R. (1993). *National innovation systems: A comparative analysis*. New York: Oxford University Press.

Nowotny, H., Scott, P., & Gibbons, M. (2001). *Re-thinking science: Knowledge and the public in an age of uncertainty*. Cambridge: Polity Press in assoc. with Blackwell Publishing.

NSTMIS. (2006). *Research and development statistics*. New Delhi: NSTMIS.

Oudshoorn, N., & Pinch, T. J. (Eds.). (2003). *How users matter: The co-construction of users and technologies: Inside technology*. Cambridge, MA: The MIT Press.

Oudshoorn, N., & Pinch, T. J. (2008). User-technology relationships: Some recent developments. In E. J. Hackett, O. Amsterdamska, M. Lynch, & J. Wajcamn (Eds.), *The handbook of science and technology studies* (pp. 541–566). Cambridge, MA: The MIT Press.

Pakrashi, S. C. (2003). Science, technology and industrial development in India. In Anon. (Ed.), *The shaping of Indian science – Indian science congress association presidential addresses – Vol. III: 1982–2003* (Vol. III, pp. 1822–1856). Hyderabad: Universities Press.

Patel, V. (2012, July). Current scenario of nano S&T in India: A comparative analysis. *Nanotech Insights, 3*(3), 48.

Patel, V., & Chander, R. V. (2012, April). Emerging trends of nanoscience and nanotechnology in India. *Nanotech Insights, 3*(2), 39–45.

Patil, S. M. (1994). *Development and applications of probes for scanning tunneling microscopy* (M Phil thesis). Department of Physics. University of Pune, Pune.

Patil, S. M., & Dharmadhikari, C. (2002). Investigation of the electrostatic forces in scanning probe microscopy at low bias voltages. *Surface and Interface Analysis, 33*, 155–158.

Patil, S. V. (2002). *Design and development of scanning force microscopic techniques for surface characterisation* (Ph D thesis). Department of Physics. University of Pune, Pune.

Philip, K., Irani, L., & Dourish, P. (2012). Postcolonial computing: A tactical survey. *Science, Technology & Human Values, 37*(1), 3–29.

PIB. (2014). *Continuation of the mission on nano science and technology in the 12th plan period*. New Delhi: Press Information Bureau, Govt. of India.

Pinch, T. J., & Bijker, W. F. (1987). The social construction of facts and artifacts: Or how the sociology of science and the sociology of technology might benefit each other. In W. E. Bijker, T. P. Hughes, & T. Pinch (Eds.), *The social construction of technological systems: New directions in the sociology and history of technology* (pp. 17–50). Cambridge, MA: The MIT Press.

Pinto, J. (2017, October 7). Let's operate in the unused toilet. *The Hindu*. Retrieved from www.thehindu.com/society/lets-operate-in-the-unused-toilet/article19818880.ece

Piston, D. W. (2012). Understand how it works. *Nature, 484*, 440–441.

Prahalad, C. K., & Mashelkar, R. A. (2010, July–August). Innovation's holy grail. *Harvard Business Review*, 10.

Prasad, A. (2006a). Beyond modern vs alternative science debate: Analysis of magnetic resonance imaging research. *Economic and Political Weekly, 41*(3), 219–227.

Prasad, A. (2006b). Scientific culture in the "Other" theater of "Modern Science": An analysis of the culture of magnetic resonance imaging research in India. *Social Studies of Science, 35*(3), 463–489.

Prasad, A. (2014). *Imperial technoscience: Transnational histories of MRI in the United States, Britain and India*. Cambridge, MA: The MIT Press.

Prasad, C. S. (2001). Towards an understanding of Gandhi's views on science. *Economic & Political Weekly*, *36*(39), 3721–3732.

Prasad, C. S. (2014). Revisiting science's social contract. *Seminar*, *654*, 55–61.

Prasad, R. (2005). Organic farming vis-a-vis modern agriculture. *Current Science*, *89*(2), 252–253.

Prasad, R. (2017, December 30). IIT-H: E-skin for motion monitoring. *The Hindu*. Chennai.

Prasad, R. (2018a, January 13). Spotting urea in the flash of an eye. *The Hindu*. Chennai.

Prasad, R. (2018b, March 10). IIT-Hyderabad's novel composite keeps tomatoes frest for 30 days. *The Hindu*. Chennai.

Prasad, R. (2018c, April 15). IASST researchers use smart bandage for faster wound healing. *The Hindu*. Chennai.

Priorities for science in India. (2015). *Nature*, *521*, 151–155.

Purie, A. (2010). Editorial. *India Today*, 1.

Purushotham, H. (2012, October–December). Transfer of nanotechnologies from R&D institutions to SMEs in India: Opportunities and challenges. *Tech Monitor*, 23–33.

Purushotham, H. (2013). CMNT initiatives: Taking nanotech to the market place. *Nano Digest*, *5*(2), 14–15.

Radjou, N., Prabhu, J., & Ahuja, S. (2012). *Jugaad innovation – think frugal, be flexible, generate breakthrough growth*. San Francisco: Jossey-Bass.

Raina, D. (2003). *Images and contexts: The historiography of science and modernity in India*. New Delhi: Oxford University Press.

Raina, D. (2007). Science since independence. In I. Pande (Ed.), *India 60* (pp. 182–195). New Delhi: HarperCollins Publishers.

Raina, D. (2014). The problem. *Seminar*, *654*, 12–14.

Raina, R. (2014). Beyond supply driven science. *Seminar*, *654*, 69–74.

Ramani, R., Sharma, K. K., Monobrullah, M., & Mohanasundaram, A. (2015). Harnessing desirable insects and managing undesirable insects: Way forward in Indian agriculture. *Current Science*, *109*(12), 2179–2180.

Rangaswamy, N., & Sambasivan, N. (2011). Cutting Chai, Jugaad and Here Pheri: Towards UbiComp for a global community. *Personal Ubiquitous Computing*, *15*(6), 553–564.

Rao, C. N. R., & Kaur, J. (2011). *Nanoworld-an introduction to nanoscience and technology*. Bangalore: Navkarnataka Publications Private Limited.

Rasmussen, N. (1997). *Picture control: The electron microscope and the transformation of biology in America, 1940–1960*. Stanford, CA: Stanford University Press.

Reddy, A. (2009). Amulya Reddy: An autobiography. In S. R. Rajan (Ed.), *Amulya Reddy – citizen scientist* (p. 373). Hyderabad: Orient BlackSwan.

Reddy, R. N. (2013). Revitalising economies of disassembly – informal recyclers, development experts adn e-waste reforms in Bangalore. *Economic and Political Weekly*, *XLVIII*(13), 62–70.

Roy, J. (2015). Train more energy economists. *Nature*, *521*, 152.

Sangvai, S. (2002). *The river and life*. Kolkata: Earthcare Publications.

Sastry, M., Kumar, A., Datar, S., Dharmadhikar, C., & Ganesh, K. M. (2001). DNA-mediated electrostatic assembly of gold nanoparticles into linear arrays by a simple drop-coating procedure. *Applied Physics Letters*, *78*(19), 2943–2945.

Sawant, S. S. (1994). *Interferometric techniques for static and dynamic characterization of piezoelectric actuators: Design, development and application* (M Phil thesis). Department of Physics. University of Pune, Pune.

SC Bans Farmers' "Jugaad." (2013). Retrieved from www.highbeam.com/doc/1G1-329924197.html

Schumpeter, J. A. (1934). *The theory of economic development: An inquiry into profits, capital, credit, interest, and the business cycle*. New Brunswick: Transaction Publishers (2012).

Schumpeter, J. A. (1939). *Business cycles: A theoretical, historical, and statistical analysis of the capitalist process* (1st ed.). New York and London: McGraw-Hill Book Company Inc.

Schumpeter, J. A. (1942). *Capitalism, socialism and democracy*. New York: Harper & Row (1975).

Sekhsaria, P. (2011). Jugaad as a materials and conceptual commons. *Common Voices*, *8*, 21–23.

Sekhsaria, P. (2013). The making of an indigenous scanning tunneling microscope. *Current Science*, *104*(9), 1152–1157.

Sekhsaria, P. (2017). How users configure producer identities – dilemas of retinoblastoma treatment in India. *Economic & Political Weekly*, *LII*(40), 57–64.

Sekhsaria, P., & Thayyil, N. (2017). Visions for India: Public participation, debate and the S&T community. *Current Science*, *113*(10), 1835–1840. doi:10.18520/cs/v113/i10/1835-1840

Sekhsaria, P., & Thayyil, N. (2019, forthcoming). 2035: Visions, technologies, democracy and the citizen. *Economic and Political Weekly*.

Sengupta, S. (2014). Travel, the New Town way. *The Telegraph*.

SET-DEV. (2009). *Knowledge Swaraj – an Indian manifesto on science and technology*. Hyderabad: SET-DEV.

Shah, A. (1998, January 19). For the bookworm. *The Indian Express*. Pune.

Shapin, S. (1988). The house of experiment in seventeenth-century England. *Isis*, *79*, 373–404.

Shapin, S. (1995). Here and everywhere: Sociology of scientific knowledge. *Annual Review of Sociology*, *21*, 289–321.

Singh, V. (2015). Improve tertiary education. *Nature*, *521*, 153.

Smith, B. L. R. & Barfield, C. E. (Eds.), *Technology, R&D, and the economy*. Washington, DC: Brookings Institution and American Enterprise Institute, pp. 140–183.

Smith, P. (2001). *Cultural theory – an introduction*. Oxford: Blackwell Publishing.

Thakur, D. (2013, June 12). The Indian way? No way? *The Hindu*.

Thomas, R. (2017). Atheism and unbelief among Indian scientists: Towards an anthropology of atheism(s). *Society and Culture in South Asia*, *3*(1), 45–67.

Thomas, R. (2018). Beyond conflict and complementarity: Science and religion in contemporary India. *Science, Technology and Society*, *21*(3), 47–64.

Thomas, R., & Geraci, R. (2018). Religious rites and scienctific communities: Ayudha Puja as "Culture" at the Indian institute of science. *Zygon: Journal of Religion and Science, 53*(1), 95–122.

TIFAC. (2015). *Technology vision 2035*. New Delhi: TIFAC.

Traweek, S. (1988). *Beamtimes and lifetimes: The world of high energy physicists*. Cambridge, MA: Harvard University Press.

TRSAS. (1986). *Press release*. Retrieved from www.nobelprize.org/nobel_prizes/physics/laureates/1986/press.html

Tyabji, L. (2015). In the eyes of the other. *Seminar, 667*, 16–19.

Tyabji, N. (2011). Jawaharlal Nehru and science and technology: 1958 scientific policy resolution reconsidered. In R. L. Hangloo (Ed.), *History of science and technology: Exploring new themes* (pp. 300–306). Jaipur: Rawat Publishers.

Umrani, R., Agrawal, D. S., & Paknikar, K. M. (2013). Anti-diabetic activity and safety assessment of ayurvedic medicine, jasada bhasma (zinc ash) in rats. *Indian Journal of Experimental Biology, 51*(10), 811–822.

UNEVOC. (1998). *Under the sun or in the shade? Jua Kali in African countries: National policy definition in technical and vocational education: Beyond the formal sector*. Berlin: UNESCO.

Valiathan, M. S. (2006). *Towards ayurvedic biology: A decadal vision document – 2006*. Bangalore: Indian Academy of Sciences.

Varma, P. K. (2004). *Being Indian – the truth about why the 21st century will be India's*. New Delhi: Penguin Books.

Varughese, S. S. (2014). The public life of expertise. *Seminar, 654*, 21–26.

Venkateswaran, S. (1994). Managing waste: Ecological, economic and social dimensions. *Economic and Political Weekly*, 2907–2911.

Visvanathan, S. (1985). *Organising for science: The making of an industrial research laboratory*. New Delhi: Oxford Univeristy Press.

Visvanathan, S. (2001, September 29). Democracy, governance and science: Strange case of the missing discipline. *Economic & Political Weekly*, 3684–3688.

von Hippel, E. (1988). *The sources of innovation*. New York: Oxford University Press.

von Hippel, E. (2005). *Democratizing innovation*. Cambridge, MA: The MIT Press.

Yehia, A. O. A. (1999). *A study of nucleation and growth of thin films using scanning tunneling microscopy and related techniques* (Ph D thesis). Department of Physics. University of Pune, Pune.

Zachariah, M., & Sooryamoorthy, R. (1994). *Science in participatory development: The achievements and dilemmas of a development movement*. London: Zed Books.

Index